A-LEVEL
GEOGRAPHY

AQA

Physical geography

David Redfern
Catherine Owen

Photo credits

pp. 97 and 101 David Redfern, p. 92 Andrew Redfern

Acknowledgements

Every effort has been made to trace all copyright holders, but if any have been inadvertently overlooked, the Publishers will be pleased to make the necessary arrangements at the first opportunity.

Although every effort has been made to ensure that website addresses are correct at time of going to press, Hodder Education cannot be held responsible for the content of any website mentioned in this book. It is sometimes possible to find a relocated web page by typing in the address of the home page for a website in the URL window of your browser.

Hachette UK's policy is to use papers that are natural, renewable and recyclable products and made from wood grown in well-managed forests and other controlled sources. The logging and manufacturing processes are expected to conform to the environmental regulations of the country of origin.

Orders: please contact Hachette UK Distribution, Hely Hutchinson Centre, Milton Road, Didcot, Oxfordshire, OX11 7HH. Telephone: +44 (0)1235 827827. Email education@hachette.co.uk Lines are open from 9 a.m. to 5 p.m., Monday to Friday. You can also order through our website: www.hoddereducation.co.uk

ISBN: 978 1 3983 2818 1

First published in 2021 by
Hodder Education,
An Hachette UK Company
Carmelite House
50 Victoria Embankment
London EC4Y 0DZ

www.hoddereducation.co.uk

Impression number   10 9 8 7 6 5 4 3 2 1

Year   2025 2024 2023 2022 2021

Cover photo © freshidea – stock.adobe.com

Typeset in India by Integra Software Serv. Ltd

Printed in India

A catalogue record for this title is available from the British Library.

Contents

■ Getting the most from this book

Exam tips
Advice on key points in the text to help you learn and recall content, avoid pitfalls, and polish your exam technique in order to boost your grade.

Knowledge check
Rapid-fire questions throughout the Content Guidance section to check your understanding.

Knowledge check answers
1 Turn to the back of the book for the Knowledge check answers.

Summaries
■ Each core topic is rounded off by a bullet-list summary for quick-check reference of what you need to know.

Exam-style questions

Commentary on the questions

Tips on what you need to do to gain full marks.

Sample student answers

Practise the questions, then look at the student answers that follow.

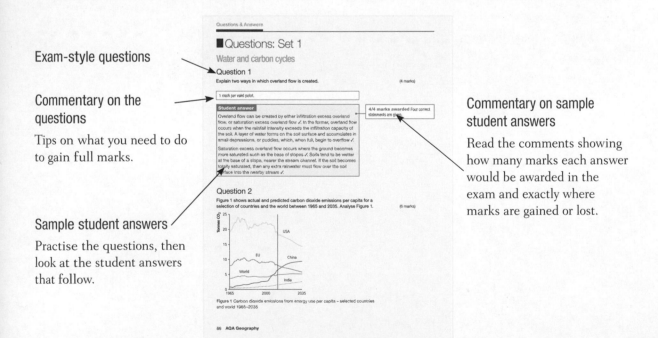

Commentary on sample student answers

Read the comments showing how many marks each answer would be awarded in the exam and exactly where marks are gained or lost.

About this book

Much of the knowledge and understanding needed for A-level geography builds on what you have learnt for GCSE geography, but with an added focus on key geographical concepts and depth of knowledge and understanding of content. This guide offers advice for the effective revision of **Physical geography**, which all students need to complete.

The first part of the A-level Paper 1 external exam paper tests your knowledge and application of aspects of Physical geography with a particular focus on Water and carbon cycles, and *one* of Hot desert systems and landscapes, Coastal systems and landscapes or Glacial systems and landscapes, and *either* Hazards or Ecosystems under stress. The Physical geography examination lasts 2 hours 30 minutes and the unit makes up 40% of the A-level award. More information on the exam papers is given in the Questions & Answers section (pages 81–85).

To be successful in this unit you have to understand:
- the key ideas of the content
- the nature of the assessment material – by reviewing and practising sample structured questions
- how to achieve a high level of performance within them

This guide has two sections:

Content Guidance – this summarises some of the key information that you need to know to be able to answer the examination questions with a high degree of accuracy and depth. In particular, the meaning of key terms is made clear and some attention is paid to providing details of case study material to help meet the spatial context requirement of the specification. Note: there is *not* a content section for Ecosystems under stress in this student guide.

Questions & Answers – this includes some sample questions similar in style to those you might expect in the exam. There are some sample student responses to these questions as well as detailed analysis, which will give further guidance in relation to what exam markers are looking for to award top marks.

The best way to use this book is to read through the relevant topic area first before practising the questions. Only refer to the answers and commentaries after you have attempted the questions.

Content Guidance

This section outlines the following areas of the AQA A-level geography specification:

- Water and carbon cycles
- Hot desert systems and landscapes
- Coastal systems and landscapes
- Glacial systems and landscapes
- Hazards

■ Water and carbon cycles

Water and carbon cycles as natural systems

In geography, two general types of **system** are recognised:

- A **closed** system is where there is transfer of energy, but not matter, between the system and its surroundings. Planet Earth is an example of such a system.
- An **open** system is where a system receives inputs and transfers outputs of energy or matter across the boundaries within it and with its surroundings. Most natural systems, such as the water cycle and carbon cycle, are open systems.

Open systems have common features that are important elements in their own right:

- **Inputs** are those elements that enter a system to be processed. They are fed into the system in order to create outputs.
- **Outputs** are the outcome of processing within the system. Outputs may be of use to the next element in the system, or they may be unintended outcomes that may not be of use.
- **Stores** (or **components**) are where amounts of energy or matter are held, and not transferred until the appropriate processes are in place to move them.
- **Transfers** or flows involve the movement of energy or matter through the system and enable inputs to become outputs. These may involve **processes** that create change.

Open systems tend to adjust themselves to flows of energy and/or matter by modifying the interrelationships between different elements of the system, so that input and output flows balance each other out, resulting in a steady state for the system, known as **dynamic equilibrium**. This kind of adjustment is called self-regulation, and much of physical geography can be understood in part as the study of self-regulating systems.

Feedback is probably one of the most important aspects of systems theory. It occurs when one element of a system changes because of an outside influence. This will upset the dynamic equilibrium, or state of balance, and affect other components in the system. **Negative feedback** is when a system acts by lessening the effect of the original change and ultimately reversing it. **Positive feedback** occurs within a system where a change causes a further, or snowball, effect, continuing or even accelerating the original change.

A **system** Any set of interrelated components that are connected together to form a working unit or unified whole.

Dynamic equilibrium The balanced state of a system when opposing forces, or inputs and outputs, are equal.

Knowledge check 1

Give a real-world example of each of (a) a negative feedback cycle and (b) a positive feedback cycle.

The water cycle

The world's water is always in the process of movement – the natural water cycle describes the continuous movement of water on, above and below the surface of the Earth. Water is always changing states between liquid, vapour and ice, with some of these processes happening within seconds and others over much longer periods of time.

The distribution and size of water stores

96% of the world's water is saline seawater; of the total freshwater, over 68% is locked up in ice and glaciers. Another 30% of freshwater is in the ground. Fresh surface water sources, such as rivers and lakes, constitute only about 1/150th of 1% of total water. Yet, rivers and lakes are the sources of most of the freshwater people use every day.

The hydrosphere (oceans)

Much more water is 'in storage' for long periods of time than is actually moving through the water cycle. The oceans hold the vast majority of all water on Earth. It is also estimated that the oceans supply about 90% of all evaporated water.

Although over the short term of hundreds of years the oceans' volumes do not change much, the amount of water in the oceans does change over the long term. During the last ice age, sea levels were lower by as much as 120 m, which allowed humans to cross over to North America from Asia at the (now underwater) Bering Strait. Similarly, the English Channel was dry. During such colder climatic periods more ice caps and glaciers form, and enough of the global water supply accumulates as ice to lower sea levels. The reverse is true during warm periods. During the last major global 'warm spell', 125,000 years ago, the seas were about 5.5 m higher than they are now.

The atmosphere

Although the atmosphere may not be a great store of water (only about 0.001% of the total Earth's water), it is the main vector that moves water around the globe. Evaporation and transpiration change liquid water into vapour, which then ascends into the atmosphere due to rising air currents. Cooler temperatures at altitude allow the vapour to condense into clouds and strong winds move the clouds around the world until the water falls as precipitation to replenish the land-based parts of the water cycle.

About 90% of water in the atmosphere is produced by evaporation from water bodies, while the other 10% comes from transpiration from plants. A very small amount of water vapour enters the atmosphere through sublimation, the process by which water changes from a solid (ice or snow) to a gas, bypassing the liquid phase. Clouds are the most visible manifestation of atmospheric water, but even clear air contains water – water in particles that are too small to be seen.

The cryosphere (ice caps, ice shelves, sea ice and glaciers)

The vast majority – almost 90% – of Earth's ice mass is in Antarctica, while the Greenland ice cap contains 10% of the total global ice mass. Collectively ice caps and glaciers cover about 10% of the Earth's surface. An ice shelf is a floating extension of land ice. Ice shelves in Antarctica cover more than 1.6 million km^2 (an area the size of Greenland), fringing 75% of the continent's coastline, and covering 11% of its total area.

Exam tip

Systems theories, involving various forms of feedback, feature several times in this book. Make sure you understand these concepts fully.

Exam tip

It is important that proportions, or percentages, of relative amounts of water are learnt and understood in this section.

Sea ice is frozen ocean water, surrounding several polar regions of the world. On average sea ice covers up to 25 million km^2, an area 2.5 times the size of Canada. One difference between sea ice and ice shelves is that sea ice is free-floating; the sea freezes and unfreezes each year, whereas ice shelves are firmly attached to the land.

Lithosphere (land-based) storage

Freshwater storage

Surface freshwater includes water courses of all sizes, from large rivers to small streams, ponds, lakes, reservoirs and freshwater wetlands. Freshwater represents only about 2.5% of all water on Earth and freshwater lakes and swamps account for a mere 0.29% of the Earth's freshwater. 20% of all surface freshwater is in one lake – Lake Baikal in Asia. Another 20% is stored in the Great Lakes of North America. Rivers hold only about 0.006% of total freshwater reserves.

Groundwater storage

Some of the precipitation that falls onto the land infiltrates deep into the ground to become groundwater. Large quantities of water are held deep underground in zones called **aquifers**. Water from these aquifers can take thousands of years to move back into the surface environment, if at all.

Factors driving change in the water cycle

Evaporation

Evaporation is the process by which water changes from a liquid to a gas or vapour.

Heat (energy) is necessary for evaporation to occur. Energy is used to break the bonds that hold water molecules together, which is why water easily evaporates at the boiling point (100°C) and evaporates much more slowly at the freezing point. Evaporation from the oceans is the primary mechanism supporting the surface-to-atmosphere portion of the water cycle.

Evapotranspiration

Evapotranspiration refers to the water lost to the atmosphere from the ground surface, evaporation from the capillary fringe of the water table, and the transpiration of groundwater by plants whose roots tap the capillary fringe of the water table. The transpiration aspect of evapotranspiration is the process by which water is lost from a plant through the stomata in its leaves.

Condensation and cloud formation

Condensation is the process by which water vapour in the air is changed into liquid water. It occurs when saturated air is cooled, usually by a rise in altitude, below the **dew point**. Condensation is crucial to the water cycle because it is responsible for the formation of clouds. These clouds may produce precipitation, which is the primary route for water to return to the Earth's surface within the water cycle.

Dew point The temperature at which a body of air at a given atmospheric pressure becomes fully saturated. If an unsaturated body of air is cooled, a critical temperature will be reached where its relative humidity becomes 100% (i.e. saturated).

Exam tip

It is likely that questions will make use of the terms lithosphere, hydrosphere, cryosphere and atmosphere. Make sure you do not get them confused.

Aquifer A permeable rock that can store and transmit water.

Evapotranspiration The combined water gain to the atmosphere and transpiration.

Knowledge check 2

Identify and explain the factors that determine transpiration rates.

Exam tip

You will have to be aware of all of these factors at a variety of scales: global (see page 7), and hill slope and drainage basin (see page 9).

Precipitation

Precipitation is water released from clouds in the form of rain, freezing rain, sleet, snow or hail. Most precipitation falls as rain. For precipitation to happen the water droplets must condense on even tinier dust, salt or smoke particles, which act as a nucleus (condensation nuclei). Water droplets may grow as a result of additional condensation of water vapour when the droplets collide. If enough collisions occur to produce a droplet with a fall velocity which exceeds the cloud updraft speed, then it will fall out of the cloud as precipitation.

Another mechanism (known as the Bergeron-Findeisen process) for producing a precipitation-sized drop is through a process which leads to the rapid growth of ice crystals at the expense of the water vapour present in a cloud. These crystals may fall as snow, or melt and fall as rain.

Sublimation and de-sublimation

Sublimation refers to the conversion between the solid and the gaseous phases of matter, with no intermediate liquid stage. It is most often used to describe the process of snow and ice changing into water vapour in the air without first melting into water. The opposite of sublimation is 'de-sublimation' (sometimes referred to as **deposition**), where water vapour changes directly into ice – such as hoar frost on trees.

Snowmelt

Runoff from snowmelt is a major component of the global movement of water, although its importance varies greatly geographically. In warmer climates it does not directly play a part in water availability. In the colder climates, however, much of the springtime runoff and flow in rivers is attributable to melting snow and ice.

Drainage basins as open systems

Other aspects of the water cycle are best examined at a smaller scale: the **drainage basin** hydrological cycle. Inputs include energy from the sun and precipitation. Outputs include evaporation and transpiration (evapotranspiration), water percolating into deep groundwater stores and runoff into the sea. Stores can take place in a number of locations – on vegetation, on the ground, in the soil and in the underlying bedrock. Transfers take place between any of these stores and ultimately into the channels of the rivers of the drainage basin. Drainage basins are bounded by high land beyond which any precipitation will fall into the adjacent drainage basin. The imaginary line that separates adjacent drainage basins is called a **watershed**.

Drainage basin terminology:

Groundwater store is water that collects underground in pore spaces in rock.

Groundwater flow is the movement of groundwater. This is the slowest transfer of water within the drainage basin and provides water for the river during drought.

Infiltration is the movement of water from the surface downwards into the soil.

Interception is the process by which precipitation is prevented from reaching the soil by leaves and branches of trees as well as by plants and grasses.

> **Exam tip**
>
> All of these processes represent transfers or flows within the water cycle system – be clear as to the direction of each of them, for example: evaporation is from land/sea to atmosphere. You may want to draw a diagram of them.

Drainage basin The catchment area from which a river system obtains its supplies of water.

Overland flow (or **surface runoff**) is the movement of water over saturated or impermeable land.

Percolation is the downward movement of water from soil to the rock below or within rock.

Runoff is all the water that flows out of a drainage basin.

Stemflow is the water that runs down the stems and trunks of plants and trees to the ground.

Throughfall is the water that drips off leaves during a rainstorm.

Throughflow is the water that moves downslope through soil.

All of these flows lead water to the nearest river. The river then transfers water by channel flow. The amount of water that leaves the drainage basin is its **runoff**. The runoff of a river is measured by its **discharge**. For any river at a given location, this is calculated by the following:

Discharge (Q) = average velocity (V) × cross sectional area (A)

The unit is cumecs, measured in $m^3 s^{-1}$.

Exam tip

You should be able to consider the factors that affect each of these flows and stores. For example, infiltration is affected by the rate of precipitation, soil type, antecedent rainfall, vegetation cover and slope.

The concept of the water balance

This refers to when inputs of precipitation (P) are balanced by outputs in the form of evapotranspiration (E) and runoff (Q) together with changes to the amounts of water held in storage within the soil and groundwater (ΔS):

$$P = E + Q + \Delta S$$

When precipitation exceeds evapotranspiration, this produces a water surplus. Water infiltrates into the soil and groundwater stores (passing below the water table). When pore spaces are saturated excess water contributes to surface runoff. When evapotranspiration is greater than precipitation, evapotranspiration demands are met by water being drawn to the surface of the soil by capillary action. Groundwater stores are depleted and runoff tends to be reduced.

Knowledge check 3

Explain *two* ways in which overland flow is created.

The storm (flood) hydrograph

A storm (flood) **hydrograph** (Figure 1) is the graph of the discharge of a river leading up to and following a 'storm' or rainfall event – its runoff variation. A hydrograph is important because it can help predict how a river might respond to a rainstorm, which in turn can help in managing the river (even more important as climate change occurs). When water is transferred to a river quickly the resultant hydrograph is described as **flashy**. This means that the river responds very quickly to the storm and often leads to **flooding**.

Hydrograph A graph of river discharge against time.

Flood A body of water that rises to overflow land that is not normally submerged.

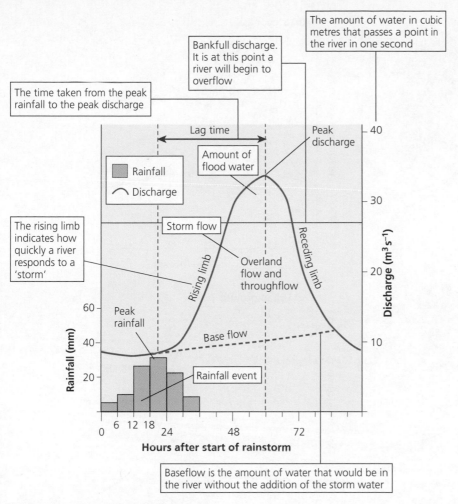

Figure 1 A storm (flood) hydrograph

A number of natural variables within a drainage basin can have an effect on the shape of the storm hydrograph (Figure 2):

- **Antecedent rainfall**: rain falling on a ground surface that is already saturated will produce a steep rising limb and a shorter lag time.
- **Snowmelt**: large amounts of water are released, greatly and rapidly increasing discharge, especially if the ground surface is still frozen, as this reduces infiltration.
- **Vegetation**: in summer, deciduous trees have more leaves so interception is higher, discharge is lower, and lag times are longer.
- **Basin shape**: water takes less time to reach the river in a circular drainage basin than in an elongated one.
- **Slope:** in steep-sided drainage basins water gets to the river more quickly than in an area of gentle slopes.
- **Geology**: permeable rocks allow percolation to occur, which slows down the rate of transfer of water to the river. Impermeable rocks allow less percolation and have greater amounts of overland flow, and hence greater discharges and shorter lag times.

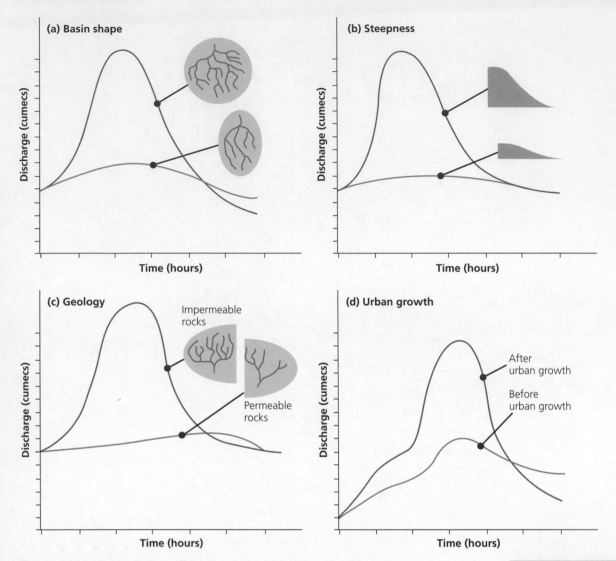

Figure 2 How key variables affect the shape of the storm hydrograph

Factors affecting change in the water cycle

Natural variations

Flash floods

These have a timeline measured in minutes or hours, but their effects can be devastating. On a hydrograph they have an extremely steep rising limb, and a very short lag time. The recession limb is equally steep and the river returns to normal flow conditions in a matter of hours.

Single event floods

These result from a single input of additional water, such as a period of cyclonic (frontal) rainfall, into the catchment. They have a timeline measured in hours and days from the onset of rainfall to the flood peak, and for the recovery back to normal conditions.

Knowledge check 4

Explain how urban growth can affect a hydrograph.

Multiple event floods

These result from a number of closely spaced single event floods in the same catchment, such that they develop a very long timescale measured in weeks, or even months, of continuous flooding. These are often the result of repeated frontal/cyclonic rainfall events, sometimes coinciding with snowmelt.

Seasonal floods

These occur where there is either seasonal rainfall or seasonal snowmelt on a massive scale. An example is the flooding associated with the Asian **monsoon**, in areas such as the Indus valley in Pakistan and the floodplains of Bangladesh.

Human factors

Land use change: deforestation

Deforestation removes water-absorbent forests, which trap and transpire rainfall, and replaces them with arable or grazing land. Consequently there will be a significant increase in both the volume of water reaching a river and the speed with which it travels. However, it has also been said that the impact of deforestation overall is overstated, with the main impact being limited to the tree felling period when tracks are being driven through the forest and heavy machinery compacts the soil causing additional overland flow. When the land reverts to scrub or pasture, runoff patterns return to their pre-deforestation state.

Farming practices: arable landscapes

It has been suggested that more emphasis on arable farming has created a greater flood risk. However, this impact varies according to the seasons. In late autumn, winter and early spring, crops are dormant and the soil is relatively bare. Rain falling on these surfaces will not be intercepted by vegetation, and hence overland flow rates are relatively high. By contrast, in late spring and summer, arable landscapes have growing and fully established crops that can intercept much greater proportions of the rainfall and thereby reduce peak flows, extending lag times.

The impact may also vary according to decision-making by farmers. For example, some farmers growing maize may clear the land after the harvest, whereas others leave the vegetation in place during the autumn to protect the soil and thereby reduce flood risk.

Water abstraction

People all over the world make great use of the water in underground aquifers. In some places they pump water out of the aquifer faster than nature replenishes it. In these cases, excessive pumping can lower the water table, below which the soil is saturated. Thus wells can 'go dry' and become useless. However, in places where the water table is close to the surface and where water can move through the aquifer at a high rate, aquifers can be replenished artificially.

Climate change

The likelihood of human-induced climate change may impact several of the factors given above. For example, in southern Asia, the seasonal monsoon rains could be affected in both magnitude and frequency, though the direction of those changes is uncertain. Some also suggest that tropical storms will be more frequent and more severe. In the UK, more intense rainstorms are likely to bring about an increased incidence of surface water flooding.

Monsoon A seasonal reversal of wind direction which brings a period of intense rain to the area affected.

Knowledge check 5

Explain how variations in precipitation can impact the water cycle.

Exam tip

Shorter examination questions will often ask you to separate out physical (natural) and human factors, but longer questions that make connections will ask you to combine them.

The carbon cycle

The distribution and size of carbon stores

Most of Earth's carbon, about 65,500 billion metric tonnes, is stored in rocks and the soil above it (the lithosphere). The remainder is in the hydrosphere (the oceans), atmosphere, the cryosphere (within the permafrost) and the biosphere (plants and animals). Carbon flows between each of these stores in a complex set of exchanges called **the carbon cycle** (Figure 3). Any change in the cycle that shifts carbon out of one store puts more carbon in the other stores – an example of dynamic equilibrium. It is now widely accepted that changes which put more carbon gases into the atmosphere result in warmer temperatures on Earth and hence the carbon cycle has a close connection to climate change.

The carbon cycle
The units of the carbon cycle are massive and confusing in much literature. This should help: 1 petagram of carbon per year (PgC/yr) = 1 gigatonne per year (1 GtC/yr) = 1 billion (10^9) tonnes = 10^{15} grams of carbon per year.

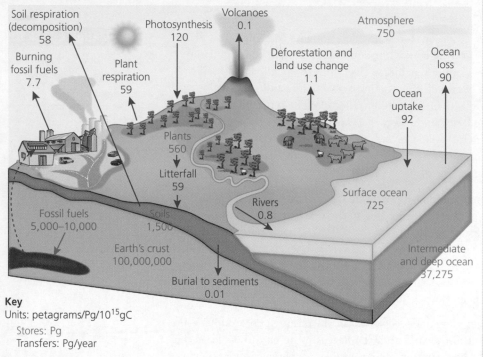

Key
Units: petagrams/Pg/10^{15}gC
   Stores: Pg
   Transfers: Pg/year

Figure 3 The carbon cycle

Factors driving change in the magnitude of carbon stores

Weathering

The movement of carbon from the atmosphere to the lithosphere begins with rain. Atmospheric carbon combines with water to form a weak carbonic acid that falls to the surface in rain. The acid dissolves rocks – a form of chemical weathering – and releases calcium, magnesium, potassium and sodium ions. Plants, through their growth, also break up surface granite, and microorganisms hasten the weathering with enzymes and organic acids in the soil coupled with the carbonic acid and carbon dioxide in the water.

Carbon sequestration in oceans and sediments

Weathered calcium and bicarbonates are then washed down to the sea by rivers and used by microscopic marine life to form shells. The ocean algae also draw down carbon dioxide from the air and when these microflora die, their remnants and the shells sink to the ocean floor and are compressed to form sediments of limestone and chalk. Carbon locked up in limestone can be stored for millions, or even hundreds of millions, of years. Carbon has also been stored in rocks where dead plant material built up faster than it could decay to become coal, oil and natural gas. Carbon is also stored in compressed clay deposits called shales.

Photosynthesis, respiration, decomposition and combustion

Plants (on land) and phytoplankton (microscopic organisms in the ocean) are key components of the carbon cycle. During **photosynthesis**, plants absorb carbon dioxide (CO_2) and sunlight to create glucose and other sugars for building plant structures. Phytoplankton also take carbon dioxide from the atmosphere by absorbing it into their cells. Using energy from the sun, both plants and plankton combine carbon dioxide and water (H_2O) to form carbohydrate (CH_2O) and oxygen. The chemical reaction looks like this:

$$CO_2 + H_2O + energy \rightarrow CH_2O + O_2$$

Four things can happen to move carbon from a plant and return it to the atmosphere, but all involve the same chemical reaction. Plants break down the sugar to get the energy they need to grow. Animals (including people) eat the plants or plankton, and break down the plant sugar to get energy (**respiration**). Plants and plankton die and decay (**decomposition**), and are eaten by bacteria, at the end of the growing season, or natural fire consumes plants (**combustion**). In each case, oxygen combines with carbohydrate to release water, carbon dioxide and energy. The basic chemical reaction looks like this (a reverse of the formula above):

$$CH_2O + O_2 \rightarrow CO_2 + H_2O + energy$$

In all four processes the carbon dioxide released in the reaction usually ends up in the atmosphere.

The carbon cycle is so tightly tied to plant life that a growing season can be seen by the way carbon dioxide fluctuates in the atmosphere. In the northern hemisphere winter, when few land plants are growing and many are decaying, atmospheric carbon dioxide concentrations climb. During the spring, when plants begin growing again, concentrations drop. It is as if the Earth is breathing.

Other forms of sequestration

Natural **sequestration** includes the following:

- **Peat bogs:** by creating new bogs, or enhancing existing ones, carbon can be sequestered.
- **Reforestation:** the planting of trees on marginal crop and pasture lands will incorporate carbon from atmospheric carbon dioxide into biomass.
- **Wetland restoration:** 14.5% of the world's soil carbon is found in wetlands, while only 6% of the world's land is composed of wetlands.

Human-induced sequestration includes the following:

- **Urea fertilisation:** fertilising the oceans with urea, a nitrogen-rich substance, which encourages phytoplankton growth.

Exam tip

Be aware of the relative sizes and rates of the various carbon cycle processes (Figure 3). It is unlikely that you will remember the numbers involved, but an understanding of relative scale is important.

Exam tip

Make sure you can define each of the terms photosynthesis, respiration, decomposition and combustion.

Sequestration The process of capture and long-term storage of atmospheric carbon dioxide by either natural or human-induced means.

- **Bio-energy with carbon capture and storage (BECCS):** where carbon is captured in power stations and stored underground.
- **Biochar:** the addition of charcoal to a soil.

Changes in the carbon cycle over time

Natural variations

Wildfires

Some define a wildfire as one caused by nature, such as a lightning strike. However, only 10% of wildfires are started this way. An alternative view is that a wildfire is one that has been started by humans, but gone out of control. It appears that forest fires can release more carbon into the atmosphere than the forest can capture and that this may be a growing problem as the number of wildfires increases. Each year wildfires (e.g. in the Amazon or in California) burn 3–4 million km^2 of the Earth's land area and release tonnes of carbon into the atmosphere as carbon dioxide. However, after a fire, new vegetation moves on to the burnt land and over time reabsorbs much of the carbon dioxide that the fire had released – an example of negative feedback.

Volcanic activity

Carbon is emitted to the atmosphere through volcanoes. Earth's land and ocean surfaces sit on several moving crustal plates. When the plates collide, one sinks beneath the other, and the rock it carries melts under the extreme heat and pressure. The heated rock recombines into silicate minerals, releasing carbon dioxide. When volcanoes erupt, they vent the gas into the atmosphere and cover the land with fresh silicate rock, to begin the cycle again. At present, volcanoes emit between 130 and 380 million metric tonnes of carbon dioxide per year.

Human impacts

Hydrocarbon fuel extraction and burning

Humans have interfered with the carbon cycle where fossil fuels have been mined from the Earth's crust and subsequently used. The age of the organisms and their resulting fossil fuels is typically millions of years. Fossil burnt fuels contain high percentages of carbon and include coal, oil and natural gas. They range from volatile materials with low carbon/hydrogen ratios such as methane, to liquid petroleum and non-volatile materials composed of almost pure carbon, such as anthracite coal.

The use of fossil fuels raises serious environmental concerns. Their burning produces around 21 Pg of carbon dioxide per year, but it is estimated that natural processes can only absorb about half of that amount, so there is a net increase of about 8 Pg of atmospheric carbon dioxide per year. Carbon dioxide is one of the greenhouse gases that enhances atmospheric heating and contributes to climate change.

Land use changes

The Intergovernmental Panel on Climate Change (IPCC) estimates that land use change, such as the conversion of forest into agricultural land, contributes a net 1.6 Gt carbon per year to the atmosphere. Various types of land use change can result in changes to carbon stores:

- conversion of natural ecosystems to permanent croplands
- conversion of natural ecosystems for shifting cultivation

Knowledge check 6

Explain how carbon exploitation (such as in power stations) and carbon capture can work together.

Knowledge check 7

Explain the role of volcanoes in the carbon cycle.

Exam tip

The impact of increased concentrations of carbon dioxide on world climate is well documented. You should be aware of what the IPCC says on this topic.

- conversion of natural ecosystems to pasture
- abandonment of croplands and pastures
- deforestation – the harvesting of timber
- wetland clearance
- establishment of tree plantations (afforestation)

Deforestation: when forests are cleared for conversion to agriculture or pasture, a very large proportion of the above-ground biomass may be burnt, releasing most of its carbon rapidly into the atmosphere. Forest clearing also accelerates the decay of dead wood and litter, as well as below-ground organic carbon. Local climate and soil conditions will determine the rates of decay; in tropical moist regions, most of the remaining biomass decomposes in less than 10 years. Some carbon or charcoal also accretes to the soil carbon pool.

Wetland clearance: when wetlands are drained for conversion to agriculture or pasture, soils become exposed to oxygen. Carbon stocks, which are resistant to decay under the anaerobic conditions prevalent in wetland soils, can then be lost by aerobic respiration.

Farming practices

Examples of impact include the following:

- Cropland soils can lose carbon as a consequence of soil disturbance such as **tillage**. Tillage increases aeration and soil temperatures, making soil aggregates more susceptible to breakdown and organic material more available for decomposition.
- Soil carbon content can be protected and even increased through alteration of tillage practices, crop rotation, residue management, reduction of soil erosion, improvement of irrigation and nutrient management.
- Heavy livestock grazing alters the ground cover and can lead to soil compaction and erosion, as well as alteration of nutrient cycles and runoff. Avoiding overgrazing can reduce these effects.
- Rice cultivation and livestock have been estimated to be the two primary sources of methane. Alteration of rice cultivation practices, livestock feed and fertiliser use are therefore potential management practices that could reduce methane sources.

The carbon budget

Key features of the **carbon budget**:

- Carbon dioxide is the single most important anthropogenic greenhouse gas (GHG) in the atmosphere, contributing 65% of **radiative forcing.**
- The current carbon budget shows a net gain of 4.4 PgC per year in the atmosphere (Figure 4).
- Rising levels of carbon dioxide and other GHGs in post-industrial times are fuelling fears of climate change through atmospheric warming.
- Atmospheric carbon dioxide reached almost 150% of the pre-industrial level in 2018, mainly due to emissions from fossil fuel combustion and cement production (caused by the production of clinker, the key constituent of cement).
- Relatively small contributions to increased carbon dioxide come from deforestation and other land use change, although the net effect of terrestrial biosphere fluxes is as a sink – 2 PgC per year.

Knowledge check 8

Explain why and how wetlands and peatlands are major carbon stores.

Tillage The mechanics of farming – the ploughing and hoeing of soil and the sowing of crops.

Exam tip

You are advised to be aware of what is being done globally to reduce the impact of deforestation and land use change. Research the work of the UN–REDD+ scheme.

Carbon budget The balance of exchanges between the four major stores of carbon.

Radiative forcing The difference of incoming solar radiation (insolation) absorbed by the Earth and the energy radiated back out into space.

■ The average increase in atmospheric carbon dioxide from 2003 to 2018 corresponded to 47% of the carbon dioxide emitted by human activity with the remaining 55% removed by the oceans and the terrestrial biosphere.

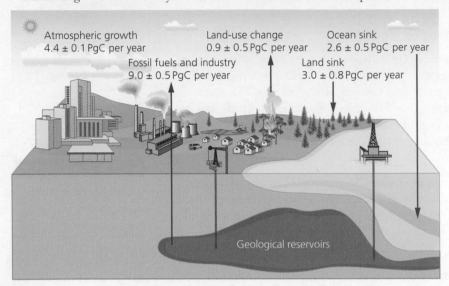

Figure 4 The carbon budget

Exam tip

Once again, numbers and proportions are important here when referring to the carbon budget. Try to remember some of these facts, and to update them as more information becomes available.

Water, carbon, climate and life on Earth

The relationship between the water cycle and the carbon cycle

Both the carbon and water stores and cycles play a key role in supporting life on Earth, largely through their influence on climate. There are clear relationships between the water and carbon cycles in the atmosphere. Furthermore, the various feedback mechanisms within and between the cycles are strongly linked to climate change. Most scientists now agree that climate change will significantly impact life on Earth.

In simple terms, the relationships between the three components, water cycle, carbon cycle and climate change, are as follows:

■ Changes in the carbon cycle are the main factors causing climate change, through the combination of the **greenhouse** and **enhanced greenhouse effects**.
■ Climate change is having, and will continue to have, an effect on the water cycle, such as increased evaporation and/or more precipitation in certain parts of the world.
■ Climate change is having, and will continue to have, an effect on the carbon cycle, such as the release of more carbon dioxide from the permafrost areas of the world as they warm.

Humans have to either **adapt** to the water cycle-related outcomes of climate change, such as increased rates of ice cap melting, flooding and drought, or **mitigate** these impacts by managing the carbon cycle, or both.

Feedbacks and climate change

Increased emissions of carbon dioxide are warming the atmosphere through the enhanced greenhouse effect. As a result, a number of positive feedback situations have arisen (see Figure 5):

Greenhouse effect
The natural processes whereby outgoing thermal radiation is trapped by atmospheric gases such as carbon dioxide and methane.

Enhanced greenhouse effect
The increased impact of greater amounts of greenhouse gases (mainly carbon dioxide and methane) caused by human activity.

Adaptation Changing lifestyles to cope with climate change.

Mitigation Reduction of the output/amount of greenhouse gases.

■ As oceans warm, more water is evaporated, which amplifies natural greenhouse warming.

■ Warm ocean water is less able to absorb carbon dioxide, resulting in more carbon dioxide remaining in the atmosphere.

■ Warmer temperatures are warming and thawing the permafrost, thereby releasing more carbon dioxide.

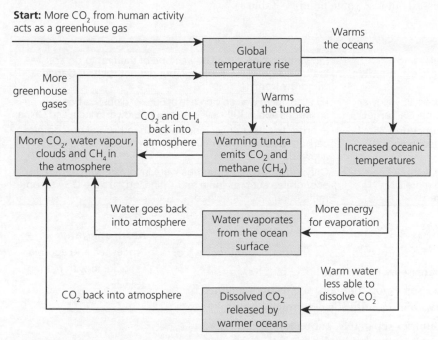

Start: More CO_2 from human activity acts as a greenhouse gas

Figure 5 Positive feedbacks of the links between carbon dioxide warming, evaporation rates and a moister atmosphere

Human interventions in the carbon cycle

There is general agreement that climate change needs to be addressed, but much less agreement on how this can be achieved. In general terms, there are two approaches: mitigation and adaptation.

Mitigation refers to the reduction in the output of GHGs and/or increasing the size and amount of GHG storage or sink sites. Examples of mitigation are:

■ setting targets to reduce greenhouse gas emissions

■ switching to renewable sources of energy

■ 'capturing' carbon emissions and/or storing or burying them (sequestration)

Adaptation refers to changing our lifestyles to cope with a new environment rather than trying to stop climate change. Examples of adaptation are:

■ developing drought-resistant crops

■ managing coastline retreat in areas vulnerable to sea-level rise

■ investing in better-quality freshwater provision to cope with higher levels of drought

You should be aware of the variety of interventions in the carbon cycle at a variety of scales: global, regional, national and local.

Global interventions include agreements such as the Kyoto Protocol and the COP21 Agreement (Paris) in 2015.

The Kyoto Protocol

The Kyoto Protocol set legally binding national targets for carbon dioxide emissions compared with 1990, and it proposed schemes to enable governments to reach these targets. The Protocol had some successes, but also some failures (Table 1).

Table 1 Evaluation of the Kyoto Protocol

Successes	Failures
The Protocol paved the way for new rules and measurements on low carbon legislation, e.g. the UK's Climate Change Act (2008).	Slow ratification – the UK was one of the first to do so; the USA signed the Protocol, but did not ratify it; Canada withdrew from it.
By 2012, carbon emissions in the EU were 22% lower than 1990 – well ahead of the initial global 5% target.	By 2015, there was an increase of 65% of global carbon emissions above 1990 levels, driven by growth in India and China.
The Clean Development Mechanism supported 75 countries in developing less polluting technology.	In order to offset emissions some countries set up complex carbon 'trading' systems, and some carbon 'sinks' were allowed.
It started a global approach to dealing with anthropogenic climate change, and more UN conferences on climate change followed.	Only industrialised countries were involved, with 'emerging' economies such as India and China left out; the USA's non-ratification has not helped.

The COP21 Agreement (Paris) 2015

A further conference took place in Paris in late 2015. Six of the main outcomes were:

- Global temperatures should not increase by more than 1.5°C by 2100.
- GHG emissions will be allowed to rise for now, but with sequestration aimed for later this century in order to keep within scientifically determined limits.
- Emissions targets will be set by countries separately, but reviewed every 5 years, with emission levels decreased meaningfully after each review.
- Wealthy countries will share science and technology relating to low GHG emissions routes to economic and social development.
- Wealthy countries will make affordable finance available for those poor nations most affected by anthropogenic climate change.
- Countries that have historically emitted a lot of GHGs (like the UK) will recognise the 'loss and damage' inflicted on poor countries because of climate change.

A **regional** approach is that of the European Union (EU) with its 20/20/20 vision which states there should be a 20% reduction in GHG emission, a commitment to 20% of energy coming from renewable sources, and a 20% increase in energy efficiency, all by 2020. The EU also offered to increase its emissions reduction to 30% by 2020 if other major emitting countries in the developed and developing worlds commit to undertake their fair share of a global emissions reduction effort.

In 2018, the EU stated it was on track to meet its 20/20/20 climate and energy targets. Data show that GHG emissions have already decreased beyond the 20% reduction target; the use of energy from renewable sources is steadily growing and is close to the 20% target; and energy efficiency levels are currently considered on track, but have decreased slightly (due to two recent cold winters).

At a **national** scale, the UK government introduced the Climate Change Act of 2008. This Act set a legally binding target for the UK to reduce GHG emissions by

Anthropogenic
Processes and actions associated with human activity.

100% compared to 1990 levels by 2050. By 2018, UK emissions were 44% below 1990 levels. It also created the independent Committee on Climate Change to advise the government and report on progress.

At a **local** scale, individuals can respond to climate change by improving home insulation, recycling, using energy more wisely (for example, using smart meters), using public transport or car sharing schemes, and calculating personal carbon footprints.

Knowledge check 9

Carbon trading systems are becoming more common around the world. Describe and explain one carbon trading system.

Case studies

You are required to consider two case studies:

1 A case study of a river catchment or catchments at a local scale to illustrate and analyse the key themes of the water cycle system and consider the impact of precipitation upon drainage basin stores and transfers. There is a strong recommendation that you carry out fieldwork, or use field data, in such an environment. You should also consider how the changes within the drainage basin impact on flooding and flood control and/or sustainable water supply. With these in mind, good examples would include upland catchments within the British Isles.

2 A case study of a tropical rainforest environment to illustrate and analyse the key themes of both the water and carbon cycle systems. You should also research the relationship of these two cycles to environmental change and human activity. With these in mind, good examples would include rainforests within the Amazon region of South America, southeast Asia and central Africa where human activity (largely deforestation and new farming practices) is contributing to change in a significant manner.

Exam tip

The exam questions on these case studies are likely to use one or more of the following words: sustainable, resilience, mitigation and adaptation. Make sure you understand these terms.

Summary

After studying this topic, you should be able to:

- understand the principles of systems in physical geography and know the meaning of terms such as inputs, outputs, flows and transfers, stores and components, and feedback mechanisms
- describe the global distribution of stores within the water cycle, and understand the factors driving change in the magnitude of these stores
- understand how drainage basins act as open systems, and the processes that cause change within them
- explain how both natural factors and human factors can affect both drainage basins and the water cycle as a whole
- describe the global distribution of stores within the carbon cycle, and understand the factors driving change in the magnitude of these stores
- explain how both natural factors and human factors can affect the carbon cycle at a variety of scales
- discuss the key roles of the carbon and water cycles in supporting life on Earth, with particular reference to climate
- evaluate the relationships between the water and carbon cycles, and how each of them contributes to climate change and has implications for life on Earth
- evaluate human interventions in the carbon cycle that are designed to influence carbon transfers and mitigate the impacts of climate change

Hot desert systems and landscapes

Deserts as natural systems

The deserts of the world are areas in which there is a substantial deficit of water, predominantly because they receive small amounts of precipitation. They have high levels of aridity; some areas of the world receive less than 100 mm, or even less, precipitation in any one year. It is this shortage of moisture, often exacerbated by high temperatures, that determines many of the characteristics of the soils, vegetation, the animals, the **landscapes** and human activities of such an area. Deserts can also be defined by the water balance – the difference between water from precipitation and losses due to evapotranspiration and changes in groundwater storage. Arid regions have an overall deficit over a year.

Desert environments are natural open systems with inputs and outputs and flows between the two.

Inputs consist of:

- energy provided by the sun through **insolation**, wind and flows of water
- sediment provided from weathering, mass movement and erosion by both wind and water
- changes resulting from human activity such as desertification

Outputs consist of:

- desert landforms both from erosion and deposition
- accumulations of sand over thousands of years
- losses of water through evapotranspiration and river flows
- sand particles blown away by winds

Stores are essentially the extensive areas of sand (ergs) that occupy about 30% of deserts, and residual areas of water (rivers and lakes), which are small in number. **Transfers** result from the actions of wind and water.

The global distribution of hot desert environments

One-third of the Earth's land surface is arid in some form (including semi-arid) – and the home to 20% of the world's population. These areas are often referred to as 'deserts', of which the Sahara, Namib, Kalahari, Atacama, Patagonian, Arabian, Thar, Mojave, Sonora and Australian deserts are the best known.

Deserts are classified in a number of ways. Some refer to warm (or hot, or mid- and low-latitude) deserts, whereas there are also deserts where, due to either high latitude or high altitude, there are winter frosts (cold deserts). Likewise, coastal deserts, such as the Atacama and Namib, have very different temperatures and levels of humidity from those deserts of continental interiors. Equally, some deserts, such as those in Arabia and Australia, have much less relief (mountains and valleys) than others. The mountain and basin deserts of southwest USA and Iran have much steeper relief and consequently have very different landscape features. Hot deserts can be classified

Landscape An expanse of land/scenery that can be seen in a single view. It covers all aspects of the view – both natural landforms and human-created features.

Making connections

The concept of the water balance was explained on page 10 in relation to drainage basins. Although the hot desert context is different, note that the concept is applicable in this environment.

Insolation The heat energy of the sun (**in**coming **sol**ar radi**ation**).

Making connections

It is important you apply the principles of systems theory (page 6), including feedback and dynamic equilibrium, to this environment.

Knowledge check 10

Describe the various types of precipitation in desert areas.

as being extremely (hyper) arid (< 50 mm precipitation per year), arid (< 250 mm precipitation per year) and semi arid (250 mm to 500 mm precipitation per year).

Characteristics of hot desert environments

Air temperatures in hot deserts are characterised by their extremes. These occur both:

- **diurnally (between day and night):** the clear skies allow both intense insolation during the day (30+°C) and rapid heat loss to space at night (in some areas below 0°C). Diurnal ranges of over 25°C are common
- **annually (throughout the year):** latitude is the most important factor in the annual range; the further away from the tropics, the greater the range

All deserts have a negative water balance, whereby evapotranspiration is in excess of precipitation. Several climatologists have attempted to devise a quantitative index, expressing the relationship between precipitation and evapotranspiration that determines aridity. In those areas of the world where there is little precipitation annually or where there is seasonal drought, the calculation of **potential evapotranspiration** is used. The best known aridity index was put forward by C. W. Thornthwaite:

Aridity index = 100 × water surplus − 60 × water deficit / Potential evapotranspiration

A value of −20 to −40 indicates a semi-arid area, and a value less than −40 indicates an arid area.

Desert vegetation and soils

Desert plants can be categorised as ephemeral, xerophytic, phreatophytic and halophytic:

- **Ephemeral** plants have a short life cycle and may form a fairly dense stand of vegetation immediately after rain. They evade drought, and when rain falls they develop vigorously and produce large numbers of flowers and fruit. The seeds then lie dormant until the next wet spell, when the desert blooms again.
- **Xerophytic** plants possess drought-resistant adaptations. Transpiration is reduced by means of dense hairs covering waxy leaf surfaces. They close their stomata to reduce water loss and either roll up or shed leaves at the beginning of the dry period. Some xerophytes – the succulents – store water in their structures. Most xerophytes have extensive shallow root networks to search for water.
- **Phreatophytic** plants have long tap roots which penetrate downwards until they reach deep sources of groundwater.
- **Halophytes** are salt-tolerant plants, common in the vast salt plains in inland desert basins.

Soils in hot deserts (**aridosols**) are coarse-textured, shallow, rocky or gravelly, with little organic matter. This is caused by the low plant productivity, which restricts the soil-building properties of microorganisms. Sometimes the accumulation of salts up through a soil progresses so far that hard surface or sub-surface crusts develop, called **duricrusts**.

Causes of aridity

Hot deserts occur in four types of locations where there is very little rain:

- latitudes dominated by dry subsiding air
- inland, far from sources of moist, maritime air

Exam tip

When asked to describe distributions or patterns of features, give an overview of common locational features rather than a list of locations.

Potential evapotranspiration
The amount of water that could be evaporated or transpired from an area, given sufficient water for this to happen.

Exam tip

Research an example of each of these four types of vegetation.

- on coasts flanked by cold ocean currents or cold upwelling ocean water
- in the rain shadow of high mountain ranges

Influence of latitude

Many of the world's largest deserts (e.g. Sahara and Australian) occur between the latitudes of 15° and 35°N and S. These are areas where the air is dry and subsiding due to the Hadley cells, and the atmospheric pressure is high for much of the year. The surface winds in such deserts are therefore generally directed outwards (trade winds), towards areas of lower atmospheric pressure, and so little moisture is brought in by surface winds. As the air over the deserts subsides it is compressed and becomes warmer, so it is able to absorb additional water vapour, reducing further the potential for rain.

Inland deserts

These are far enough inland to be away from the influence of moist maritime air masses. Rainfall decreases rapidly away from the coast in all parts of the world except those close to the equator. The drying influence of a large land mass is referred to as continentality, and this applies to all big deserts, including the great tropical deserts of Arabia, Australia and the Sahara. In these hot, tropical deserts, the dryness caused by latitude is accentuated by continentality. The deserts of central Asia, including the Taklamakan and Gobi deserts of China and Mongolia, are in the interior of mid-latitude continental regions far from the oceans.

Ocean currents

The presence close offshore of cold upwelling water or a cold **ocean current** can cause coastal aridity (due to reduced evaporation) in tropical and even in equatorial latitudes, such as the arid Horn of Africa, flanked by the cold Somali current. In fact, the western borders of all the great tropical or trade-wind deserts in both hemispheres are washed by cool ocean currents associated with the oceanic circulation cells which flow clockwise in the northern and anticlockwise in the southern hemisphere. These include the Atacama and Kalahari deserts.

Rain shadow effect

Wherever ranges of hills or mountains lie close to the coast, forming a physical barrier to onshore winds, the incoming moist maritime air will be forced upwards. Moist air becomes cooler as it rises and expands, attains vapour saturation, and sheds its condensed water vapour as rain or snow. The air then passes over the coastal ranges and flows downhill, becoming warmer and drier. The area inland of the coastal ranges, such as Patagonia in South America, is described as being in the rain shadow of the ranges – the air has already shed its moisture before passing over this land, which therefore gets little rain.

Systems and processes
Sediment sources, cells and budgets
Sediment sources

There is a variety of sources of sediment in hot deserts. It may be derived from:
- the weathering of the underlying rock

> **Knowledge check 11**
>
> Define the term Hadley cells and explain how they are formed.

> **Ocean currents** Large-scale movements of water within the oceans that are part of the process of the transfer of heat from the equator to the North and South Poles.

> **Knowledge check 12**
>
> Winds known locally as the Chinook are often associated with the rain shadow effect. Describe and explain these winds.

- fluvial deposits – where past and present rivers have brought sediment into desert regions
- aeolian deposits – where wind has transported and deposited material from beyond the desert margins

Sediment cells and budgets

Deserts can be described as being dominated by either erosion or deposition. Areas dominated by erosion are a source of sediment and the system has a **net sediment loss**. In areas dominated by the transportation and deposition of sediment, the system has a **net sediment gain**. This classification is somewhat simplistic as most deserts have areas of gain and loss. Furthermore, large deserts can be significant sources of sediment on a global scale. Satellite imagery often shows large clouds of dust (loess) being blown thousands of miles around the world. For example, the large loess deposits of central Asia and northern China, and central Canada, although not formed in hot deserts, were certainly transported from adjacent deserts.

Weathering and mass movement

Mechanical weathering, involving the disintegration of rocks without any chemical change, takes place in hot deserts due to two main factors:

- the high rates of insolation
- the action of salt

Insolation weathering is the rupturing of rocks and minerals primarily as a result of large daily temperature changes which lead to temperature gradients within the rock mass. These are manifested in a number of ways:

- **Exfoliation (onion skin weathering)** is most noticeable on rocks with few joints or bedding planes (e.g. granite, massive sandstone). It occurs because the surface heat does not penetrate very deeply into these rocks. The outer layers expand and contract daily, but the interior does not. This differential leads to weaknesses parallel to the surface. Layers peel off the rock in an onion skin-like way.
- **Granular disintegration** occurs in rocks that are comprised of minerals of different colours. Darker minerals absorb more heat than lighter ones (e.g. black mica expands more than grey quartz) and so the rock breaks up grain by grain to produce sand-sized material.
- Those rocks that are microcrystalline (e.g. basalt) are subject to shattering (**thermal fracturing**) by constant diurnal expansion and contraction due to heating and cooling respectively.
- **Block separation** occurs predominently on well-jointed and bedded limestones, where the rock breaks up into blocks along these weaknesses.
- **Frost shattering** can occur in deserts where there is some free water and the nighttime temperatures fall below zero. The water gets into fractures in the rock, freezes at night and so expands. Continual freezing and thawing will result in fragments breaking off.

Geographers now believe that these mechanical forms of weathering are encouraged by the introduction of water, and particularly so when water moves up through groundwater bringing with it weathered salts from below ground. These may enhance mechanical forms of weathering, but there may also be some chemical changes.

Exam tip

When describing different types of weathering, link them to specific rock types, as not all rocks are weathered in the same way.

Salt weathering operates in two ways:

- When a solution containing salts is either cooled or evaporated, salt crystals will form and pressures accompanying this crystallisation can be great enough to exceed the tensile strength of the rocks in which the solution was contained.
- Salt minerals expand when water is added to them – a process called hydration. In the case of sodium sulfate and sodium carbonate the expansion may be as great as 300%. If these salts are in rocks, the pressures generated can break them.

With the general lack of water in the desert environment, most forms of mass movement rely on the natural force of gravity. Rock falls and occasional rockslides are the main form of movement whereby weathered or fragmented rock falls from high up a slope down on to the ground below.

The role of the wind (Aeolian processes)

As deserts are so dry and vegetation cover so limited, there is little to protect the desert surface against the action of the wind. Wind is able to erode desert surfaces. It does this in two ways:

- **Deflation**: loose, fine material is picked up by the wind and is transported and deposited elsewhere. It can create depressions in the desert floor (deflation hollows) e.g. the Qattara depression in Egypt, or it can remove fine sand from a surface, leaving behind coarser stones that blanket the surface to form a reg or desert pavement.
- **Abrasion**: a sand-blasting effect, where fine material carried in the wind is blown at rocks. Ventifacts are rocks lying on the floor of deserts that have been shaped by the wind-driven sand. They usually have sharp edges and smooth sides. Other common landforms include zeugen and yardang (Figure 6). The coarsest sand grain material, which has the greatest effect, cannot be lifted more than 1.5 m off the ground. This results in undercutting and fluting. Zeugen form where there are large areas of horizontal rock layers with vertical jointing. Abrasion erodes the weak joints and then softer layers are eroded underneath. Thus long ridges develop with a protective cap rock. Yardang occur when the layers of rock lie at a steep angle to the surface and parallel to the prevailing wind. The less resistant rock layers are eroded more rapidly than the resistant ones, producing long rock ridges. These can be several kilometres long and hundreds of metres high.

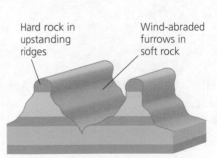
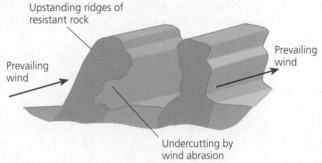

Figure 6 Zeugen and yardang

Wind is able to **transport** material in three ways:

- **Saltation**: medium-sized grains of sand (0.15–0.25 mm diameter) move in a series of hops across the surface. The wind picks up a grain to a height of a few centimetres, carries it a short distance, and then drops it. Where the grain falls it can dislodge other grains that are then picked up, and the process continues.
- **Suspension**: high-velocity winds can lift and carry fine silt and clay (< 0.15 mm diameter) high into the atmosphere. This can often be taken away from the desert area entirely. Saharan dust has reached places as far away as the UK.
- **Surface creep**: coarser grains of sand (> 0.25 mm diameter) are rolled across the surface.

Landforms resulting from deposition

Deposition occurs when the wind does not have the energy to move desert material. These deposits can then be shaped by the wind. Only about one-third of the world's deserts are covered by wind-blown sand. However, great ergs (seas of sand) can form some of the most striking landscapes seen on Earth.

Sand dunes are formed by re-working of large deposits of sand by wind. Loose sand is blown up the windward side of a dune. The sand particles then fall to rest on the downwind side (usually at an angle of about 34°), while more are blown up from the windward side. In this way a dune moves gradually downwind. The geometric forms of dunes are varied and depend on the supply of sand, the nature of the wind regime, the extent of vegetation cover and the shape of the ground surface.

The typical crescent-shaped dune, with two horns pointing downwind, is called a **barchan**. It may be up to 30 m high and 400 m across, and may move downstream at a speed of 15 m a year. It is formed by there being a dominant prevailing wind direction with it being orientated with its axis at right angles to the wind direction. The horns extend downwind as there is less sand at the edges.

Seif dunes are longitudinal and lie parallel to the prevailing wind direction. They can be very long (> 100 km) and can reach 30 m in height. They are commonly 200–500 m apart. Some suggest that they are formed in areas where there is a seasonal change in wind direction that connects crescent (barchan) dunes. Others state that they can be attributed to some regularity in patterns of turbulence in the wind.

The role of water

Present and past river action are also important in the moulding of desert landscapes. Although rainfall quantities are low overall, substantial amounts of rainfall may occur from time to time. Moreover, many desert surfaces have a number of characteristics that enable them to generate considerable runoff from quite low rainfall intensities:

- Limited vegetation cover provides little organic matter on the surface to absorb water.
- The sparseness of the vegetation means that humus levels in the soil are low, and combined with minimal disturbance by plant roots, this makes the soil dense and compact.
- As there is virtually no plant cover to intercept rainfall, rain is able to beat down on the soil surface with maximum force.

Exam tip

A good way to demonstrate that you know what each of these landforms looks like is to draw a sketch of it. Give it a go!

Exam tip

You will be asked to describe landforms. Refer to size, shape, nature of sediments and field relationship (i.e. where the landform lies in relation to the landscape).

Knowledge check 14

Another type of dune is a star dune. Describe its characteristics and explain its formation.

Making connections

The principles of drainage basin hydrology were explained on page 9. Although the hot desert context is different, note that the principles are also applicable in this environment.

■ Fine particles, unbound by vegetation, are redistributed by splash to lodge in pore spaces and create a surface of much reduced permeability.

Consequently, as a result of all these factors, infiltration rates are very low, and overland flow is highly likely. **Channel flash floods** and sheetfloods are common in deserts following short periods of intense rainfall which cannot infiltrate. They can be devastating in their impact despite being short lived.

There is a variety of water sources in desert areas:

■ **Exogenous** rivers are large perennial rivers that originate outside the desert, for example the River Nile's main tributary, the Blue Nile, which originates in the Ethiopian Highlands.
■ **Endoreic** rivers are also perennial and flow into an inland drainage basin in the desert. For example, the River Jordan flows into the Dead Sea which, at 422 m below sea level, has no outlet.
■ **Ephemeral** rivers are created by heavy rainstorms in a desert. The rain cannot infiltrate into the baked desert surface and flows over land as sheet flow or in what are normally dry valleys (wadis). After the storm, the water quickly evaporates or soaks into the wadi floor.

The action of water has produced some distinctive landforms.

Wadis are dry stream channels or valleys found in deserts. They are formed either by ephemeral or endoreic rivers. When the rivers are flowing, they can carry huge amounts of sediment. They are able to erode vertically to form steep sides. When the rivers dry up this sediment is deposited on the floor of the wadi, to give it its typical flat floor. In some parts of the Sahara there are extensive systems of wadis which could not have been created by present-day water erosion. It is likely that they were excavated by water erosion at a time when the rainfall was heavier than at present. Some believe they were formed during the ice ages – the periods of advancing ice sheets in Europe coincided with periods of heavier rainfall (pluvials) in North Africa.

Alluvial fans (also known as **bahadas** or **bajadas**) are cones of sediment that occur between mountain fronts and low-lying plains. Their size is variable – small ones may have a radius of only a few tens of metres while larger ones may be more than 20 km across, and 300 m thick at the apex. The larger ones consist of a gently sloping plain stretching from the mountain edge into the desert. They form where rivers emerge from a confined upper mountain valley, or wadi (or canyon). At such a point the river can spread out, decreasing its velocity so that deposition occurs from a sediment-rich flow.

Other aspects of arid landscape development
Desert slopes

Slope profiles in desert areas are often visually dramatic. Their form is often analysed in terms of an idealised profile: namely, a cliff or mountain front, a straight segment, and a **pediment**. Pediments may coalesce and cover extensive areas and from them may rise residual outliers called **mesas** and **buttes**.

Sheetfloods These floods remove thin layers of surface material evenly from an extensive area of gently sloping land, by broad continuous sheets of running water rather than by streams flowing in channels.

Knowledge check 15

Badlands are features of semi-arid landscapes in the USA. Describe their characteristics and explain their formation.

Exam tip

Again, you will be asked to describe landforms. Refer to size, shape, nature of sediments and field relationship (i.e. where the landform lies in relation to the landscape).

Pediments are extensive low-angled (1–7°) rock surfaces at the base of desert mountain fronts or cliffs. They may have a thin veneer of rock debris. Their origin is uncertain, but it is thought that a combination of the retreat of the mountain front by weathering accompanied by removal of debris by sheets of water or sheetfloods formed them.

Rising above the pediments are often steep-sided rocky hills called **inselbergs**, formed as a result of gradual slope retreat produced by weathering and erosion at the break of the slope. They can be up to 600 m high and form spectacular features in many deserts. Inselbergs occur in many parts of the world, including Africa and Australia; Uluru is perhaps the most famous such rock in the world.

Mesas and **buttes** are particular forms of inselbergs, associated with the USA. They are formed in areas of sedimentary rocks with horizontal bedding planes. Water has eroded the land around them, leaving some resistant portions, with a resistant cap rock standing out from the surface. Mesas are steep-sided and flat-topped. Buttes are similar, but are much more eroded and are pillar-like in appearance. Their lower slopes are often covered in scree – the result of weathering.

Away from the steep relief of these mountains, sometimes shallow salt lakes, called **playas**, develop in lowland basins. Sometimes the water is evaporated from these lakes to leave extensive salt pans. For example, the extensive flats known as Lake Bonneville can be found in the vicinity of the Great Salt Lake in Utah, USA.

Desertification

> Desertification is the degradation of land in arid, semi-arid, and dry sub-humid areas. It is caused primarily by human activities and climatic variations. Desertification does not refer to the expansion of existing deserts. It occurs because dryland ecosystems, which cover over one-third of the world's land area, are extremely vulnerable to over-exploitation and inappropriate land use.

(UNFAO)

The United Nations Convention to Combat Desertification (UNCCD) also defines desertification as 'land degradation in arid, semi-arid and dry sub-humid areas resulting from various factors, including climatic variations and human activities'. In turn, land degradation is defined as 'the reduction or loss of the biological or economic productivity of drylands'.

The main **physical cause** is climate change. In the vulnerable areas, the following changes are taking place:

- There is less rainfall, and what there is, is less reliable. This means farmers find it difficult to plan ahead. There is an increase in the frequency and intensity of droughts.
- There are higher temperatures. This increases evaporation, reduces condensation and leads to less rainfall.
- As there is less rainfall, rivers dry up and the water table falls.

Knowledge check 16

Make three lists: desert landforms created by wind; desert landforms created by water; and distinctive desert slope landforms.

Making connections

There is a clear link here to the sections on climate change within the chapter on the carbon cycle (page 18).

The prospects of even less rainfall in the future have featured in all the IPCC Assessment Reports (ARs) produced so far.

There are several **human causes**:

- The main human cause is population growth. It is estimated that over 1 billion people live in areas at risk. The numbers of these people are increasing because of high birth rates and an influx of refugees from conflict or drought-hit areas. This population growth has led to increasing livestock numbers, which in turn leads to overgrazing.
- Farming of marginal or low-quality land is causing desertification worldwide. Farmers are clearing marginal land, and overusing it, which takes away the richness in the soil – they are not letting it have a fallow period thereby allowing it to replenish itself before farming.
- Deforestation is causing desertification to occur. People are cutting down trees to use them as a source of fuel for the growing populations. Once all these trees are cut down there is nothing to protect the soil. Therefore, it turns to dust and is blown away by the wind.
- Unsuitable irrigation systems are commonly used in poorer areas. Farmers often use canal irrigation and other poor techniques because of the lack of water. This type of irrigation causes a build-up of salt in the soil, which makes it difficult for plants to grow due to the salinity.
- The changing nature of world trade in food has led to an increase in the growing of cash crops for markets in the more developed world (such as coffee in Ethiopia). This means that the best land is used for these crops and unsuitable marginal land is used for growing food crops.
- Several areas suffering from the other causes of desertification are often caught up in political instability and conflict. These cause poor land management practices to come to the fore, or abandonment.

All of these factors combine and lead to further reductions in vegetation on the land. Soil is then exposed to the wind and rain, which in turn leads to soil erosion because the root mat that holds the soil together is gone. This loss of soil then leads to desertification – an example of how physical and human feedback processes operate in a cumulative, negative and damaging way.

Managing the world's dryland environments is perhaps one of the most challenging and pressing development problems of today. Long-term strategies for eradicating poverty in the world's drylands will only succeed when the natural resources on which the people who live there depend are used sustainably. Programmes for protecting the dryland environment will accomplish their aims only when they also address the day-to-day pressures of poverty.

In general terms, there are two approaches to managing the drylands:

- To mitigate against the drought. Drought cannot be prevented, although scientists have experimented with cloud seeding using silver iodide pellets to bring on rainstorms. Dealing with drought mainly involves setting up water storage schemes and increasing community preparedness.
- To adapt to the occurrence of the drought and modify **vulnerability**, thereby increasing **resilience**. Scientists use satellite imagery to measure the progress of

---

Exam tip

You could research desertification in the latest IPCC report, *AR5*. See what it says about the prospects of the case study area you examine in detail.

Knowledge check 17

Explain how overgrazing leads to desertification.

Making connections

This is an example of where the principles of systems theory, including feedback and dynamic equilibrium (page 6), apply to this environment.

Vulnerability A set of conditions and processes resulting from physical, social, economic and environmental factors, which increase the susceptibility of a community to the impact of hazards.

Resilience The ability of a system, community or society exposed to hazards to resist, absorb, accommodate to and recover from the effects of a hazard in a timely and efficient manner.

the rains in order to predict drought areas before people begin to suffer due to crop failure. This provides some accuracy to warnings, but often aid and government agencies lack the resources to respond in time. Longer-term drought aid in the form of irrigation schemes and education for farmers on water conservation techniques is perhaps more useful. Another more specific activity is to engage in community-based agroforestry schemes to try and maintain the natural environment for all users.

Case studies

You are required to consider two case studies:

1 A case study of a hot desert environment, or a setting similar to it, to illustrate and analyse some of the key themes set out above and to engage with field data. It may be that you have been to a hot desert area where processes associated with wind and water can be examined at first hand. However, it is also recognised that a similar setting may be used that is *not* in a hot desert environment, such as an area of coastal dunes in the UK. In this latter scenario, it is most likely that you will have studied the action of wind (known as aeolian processes, see page 26). The key aspect is that you make use of data collection methods based on either moving sand or water in either area of study.

2 A case study of a setting in a desert at a local scale to illustrate and analyse key themes of desertification. The chosen area could be in any area of the world, such as the Sahel of Africa, the southwest USA or fringe areas of the Australian desert. You should look at its causes and impacts, and assess the implications of these impacts on sustainable development. You should also examine and evaluate the human responses to these causes and impacts with a view to assessing resilience, and the extent to which people can mitigate or adapt to the conditions there.

Summary

After studying this topic, you should be able to:

- explain how deserts operate as natural systems, and describe their main features such as global distribution, climate, vegetation and soils
- explain the variety of ways in which deserts are caused
- discuss the range of geomorphological processes that operate in desert environments including weathering, erosion, transportation and deposition by the actions of wind and water
- describe the variety of landscapes that have been created in desert environments over time, largely due to the actions of wind and water
- discuss the concept of desertification, its causes and impacts
- evaluate the impact of change in areas subject to desertification, such as climate change, and assess the implications for the futures of people living in these areas

Exam tip

Desertification illustrates physical and human processes operating together, often involving feedback mechanisms. Draw a diagram to help you clarify these links.

Exam tip

Exam questions on the second case study are likely to use one or more of the following words: sustainable, resilience, mitigation and adaptation. Make sure you understand these terms.

Coastal systems and landscapes

Coasts as natural systems

Coasts are important elements of the natural **landscape**. This is a result of their history of sea-level change (some of which is taking place currently), the geological structures that lie behind them, the sediments that are available to make their beaches and the waves and tides that mould them. There is a variety of processes (erosion, deposition and sea-level change) operating along coastlines. These develop a variety of landforms which, when assembled, result in the varying landscapes that have developed along coastlines.

There is also a variety of strategies as to how coastlines should be protected and/or managed, if indeed they should be at all. Here, **sustainability** is a key criterion that should be considered.

Coastal environments are natural, open systems with inputs and outputs and flows between the two.

Inputs consist of:

- energy provided by waves, winds, tides and currents. This is increased at times of storms and storm surges
- sediment provided from the erosion of coastlines by waves, as well as that brought by rivers. Weathering and mass movement also contribute material from cliff faces
- changes in sea level – as sea levels rise with climate change, more energy is exerted on a coastline

Outputs consist of:

- coastal landforms, both erosional and depositional
- accumulations of sediment above the tidal limit (e.g. dunes)
- loss of wave energy through processes such as refraction

Stores in the coastal system refer to the water in the sea, and sand/shingle on beaches.

Transfers result from the actions of wind and waves. An example is the process of longshore drift.

Systems and processes

Waves

Waves are caused by the wind blowing over the surface of the sea. As the wind drags over the surface of the water, friction causes a disturbance and forms waves. Waves at sea follow an orbital movement and objects on the water do not travel forward. The resultant up/down movement at sea is called the swell. However, when a wave reaches shallow water, the movement of the base of the wave is slowed by friction with the seabed, and the wave spills forward as a breaker. Water rushes up a beach as swash, before drawing back to the sea as backwash.

Wave energy is controlled by:

- the force of the wind and its direction
- the duration of wind
- the fetch – the longer the fetch, the more energy waves possess

Landscape An expanse of land/scenery that can be seen in a single view. It covers all aspects of the view, both natural landforms and human-created landforms.

Sustainability Meeting the needs of today without compromising the ability of future generations to meet their own targets.

Exam tip

Note that this topic should be illustrated with examples from both the UK and beyond the UK.

Making connections

It is important you apply the principles of systems theory (page 6), including feedback and dynamic equilibrium, to this environment.

It is common to classify waves as being either constructive or destructive:

- **Constructive** waves construct or build beaches and are usually the product of distant weather systems. They have longer wavelengths, lower height and are less frequent, at 6–8 per minute. Swash is greater than backwash, so they add to beach materials, giving rise to a gently sloping beach. The upper part of such a beach is marked by a series of small ridges called berms, each representing the highest point the waves have reached at a previous high tide.

- **Destructive** waves have a shorter wavelength, a greater height and are more frequent, at 10–14 per minute. The backwash is greater than the swash so that sediment is dragged offshore. This creates a steeper beach profile initially, though over time, the beach will flatten as material is drawn backwards. Destructive waves also create shingle ridges at the back of a beach, known as storm beaches.

Currents and tides

On sandy beaches with large waves, the underwater part of the beach is often characterised by alternating shallow and deep sections known as ridges and runnels. These channels can either run parallel to the beach or at right angles to it. Strong **currents** can be present in these channels and when their flow speed is high, they can be dangerous to swimmers.

Tides are caused by the gravitational pull of the moon, and less so the sun, usually causing two high tides and two low tides each day (Figure 7). In the UK, the interval between each high tide is 12 hours 25 minutes. Tidal range refers to the difference between the height of the high and low tides. Where there is a larger difference, as around the west coast of Britain, it is said to be macrotidal; erosion is greater and tidal bores can develop in estuaries. The Severn Estuary has a tidal range of 15 m but the world's largest tidal range is the Bay of Fundy in Canada at 17 m.

Knowledge check 18

Define two key terms associated with coasts: fetch and tides.

Exam tip

When asked to compare or contrast features, such as different types of wave, make sure you make clear comparative statements rather than separate statements.

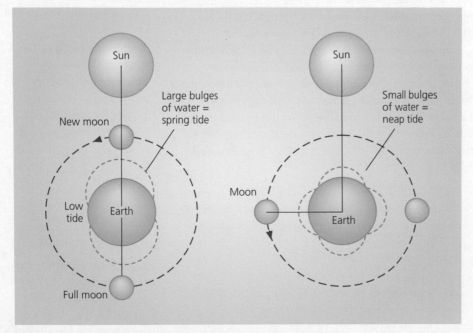

Figure 7 Spring and neap tides

Spring tides occur when the Sun, the Moon and the Earth are in a straight line, which happens twice a lunar month when the gravitational pull is strongest. It causes the high tide to be at its highest point and the low tide is at its lowest point. **Neap tides** happen during the weeks in the lunar month when the Earth, the Moon and the Sun form a right angle, lessening the overall gravitational pull and giving a lower tidal range.

Tides are important influences on a coastline. If there is a small tidal range, the power of the waves (and hence erosion) is concentrated on a relatively narrow element of coastline, such as a cliff base. If there is a large tidal range, large tracts of beach can become exposed at low tide. When this sand dries, onshore winds can blow the sand to create sand dunes inland.

Knowledge check 19

Describe the nature of strong currents known as rip currents.

Low-energy and high-energy coasts

High-energy coastlines are those where wave power is strong for a greater part of the year, e.g. the western coast of the British Isles. The prevailing and dominant wind direction on these coasts is westerly and they face the direction of the longest fetch. The maximum-recorded wave height on western coasts is therefore greater than that on eastern coasts. Waves up to 30 m have been recorded on the west coast of Ireland.

Many estuaries, inlets and sheltered bays are **low-energy coastlines,** where wave heights are considerably lower. Here waves spread outwards and energy is dissipated, leading to the deposition of transported material. Enclosed seas also contain low-energy environments. The Baltic Sea contains some of the longest depositional landforms in the world because of its sheltered water and low tidal range.

Sediment sources and cells

Sediment sources include rivers, the seabed, erosion of the coastline, shell material and movement of materials along the coastline. **Sediment** (or **littoral**) **cells** are distinct areas of coast separated by deep water or headlands within which material is moved but the sediment inputs and outputs are in balance – a type of sediment budget. In the UK, the Department for Environment, Food and Rural Affairs (DEFRA) defines a sediment cell as 'a length of coastline and its associated nearshore area within which the movement of coarse sediment (sand and shingle) is largely self-contained. Interruptions to the movement of sand and shingle within one cell should not affect beaches in a neighbouring sediment cell'. In theory, they are regarded as closed cells (from which nothing is net gained or lost), but in reality finer material does move into neighbouring cells.

Sediment budget The relationship between accretion and erosion, which can be used to predict the changing shape of a coastline over time.

Eleven major sediment cells have been identified for England and Wales as basic units for coastal management. Sediment cell theory is a key component of Shoreline Management Plans, the authors of which decide on future strategies of coastal management. (See page 43.)

Weathering

Physical weathering involves the breakdown of rocks into smaller fragments through mechanical processes. Two types of physical breakdown that are common on coasts are frost shattering weathering and salt weathering. These are examples of **subaerial weathering**.

Subaerial weathering The collective name for weathering processes on the Earth's surface (literally at the base of the atmosphere).

- **Frost shattering** (or freeze–thaw) takes place in rocks that contain crevices and **joints**, and where temperatures fluctuate around 0°C. Water enters the joints, and during cold nights freezes. As ice occupies around 9% more volume that water, it exerts pressure within the joint. This alternating freeze–thaw process slowly widens the joints, eventually causing bits to break off from the main body of rock. It leads to the formation of **scree**.

- **Salt weathering** takes place when a rock becomes saturated with water containing salt, as in coastal environments. Some of the salt crystallises and begins to exert pressure on the rock, as the salt crystals are larger than the spaces in which they are being formed. As with frost shattering, the process repeats over time, and causes the disintegration of rock.

Chemical weathering involves the decay or decomposition of rock in situ. It usually takes place in the presence of water, which acts as a dilute acid. The end products of chemical weathering are either soluble, or they have a different volume – usually bigger – than the mineral they replace. The rate of chemical weathering tends to increase with rising temperature and humidity levels, except in the action of carbonic acid (carbonation), where lower temperatures produce greater rates of weathering on limestones. Chemical weathering can also occur from the action of dilute acids resulting from both atmospheric pollution (sulfuric acid), and the decay of plants and animals (organic acids).

Another means by which material reaches the sea is through **runoff** – the flow of water overland either as rills in small channels, or as rivers and streams.

Mass movement

The rate of **mass movement** depends on the degree of cohesion of weathered material (**regolith**), the steepness of the slope down which the movement takes place, and the amount of water contained in the material. A large amount of water adds weight to the mass, but more importantly lubricates the plane along which movement can take place.

There are various forms of mass movement that affect coastlines:

- **Creep:** the slow downhill movement of soil and other material. It operates on slopes steeper than 6° and evidence is shown by small **terracettes** on a hillside.

- **Earthflows:** when weathered material becomes saturated, internal friction between the particles is reduced and stress can cause the debris to move under gravity. This occurs on slopes as gentle as 5° once mobile, but usually needs a slope of about 10° to initiate movement. Such flows are faster than creep.

- **Mudflows:** more rapid flows that can occur on relatively low slope angles compared with earthflows. They occur in areas which experience torrential rain falling on ground which has limited protection from vegetation cover. This allows the regolith to become saturated, increases the pore water pressure in the debris and reduces the frictional resistance between particles.

- **Rock falls:** when erosion is concentrated at the base of a cliff, it will become unstable and collapse into the sea.

- **Landslides (landslips):** occur when rocks and/or regolith have bedding planes or layers, and material in one plane/layer becomes very wet and over-lubricated. The added weight from the water will cause the plane/layer to slip under gravity downwards over the underlying layers.

> **Exam tip**
>
> Ensure you know the difference between weathering and erosion processes.

> **Mass movement** The downslope movement of weathered material under the influence of gravity.
>
> **Regolith** The collective name for all of the material weathering produces.

- **Slumping:** when saturated material moves suddenly, resulting in whole sections of cliffs moving down towards a beach. This happens particularly where softer material overlies strata which are far more resistant. The slip plane is often concave, producing a rotational movement.

Erosion, transportation and deposition

Coastal erosion operates through a variety of processes:

- **Abrasion (corrasion):** when the material waves carry (load) is used as ammunition to wear away rocks on a cliff or a wave-cut platform when repeatedly thrown or rubbed against these landforms. Where abrasion is targeted at specific areas, such as notches or caves, it is referred to as **quarrying**.
- **Attrition:** the process by which loose rocks are broken down into smaller and more rounded pebbles, which are then used in abrasion.
- **Hydraulic action:** where a wave breaking against rocks traps air into cracks in the rock under pressure, which is then released suddenly as the wave retreats. This causes stress in the rock, which develops more cracks, allowing the rock to break up more easily. Hydraulic action also includes **pounding** – the sheer weight and force of water pushing against a cliff face causing it to weaken.
- **Cavitation** occurs when air bubbles trapped in the fast-moving water collapse, causing shock waves to break against the rocks under the water. Repeated shocks of this nature weaken the rock.

Exam tip

Despite what it may say in the specification, solution is *not* a major form of erosion by the sea – it is too alkaline.

The rate of coastal erosion is governed by several factors:

- **Geology:** harder rock, for example granite, is more difficult to erode than softer material like boulder clay. Some more resistant rock can be eroded along joints or cracks, and limestones are prone to weathering. The structure and dip of rocks also affect erosion. If rocks dip inland or are horizontally bedded, steep cliffs form, whereas those rocks which dip seaward produce gentler slopes.
- **Coastal shape:** softer rock is eroded to form bays, with the harder rocks forming headlands. Wave refraction is concentrated on headlands, causing erosion here, whereas in the bays the waves spread over a wider area and their energy is dissipated, causing more deposition.
- **Wave steepness:** steeper, high-energy waves have more power to erode.
- **Wave breaking point:** waves that break at the foot of a cliff have more energy to erode.
- **Fetch:** waves that have travelled a long distance have more energy.
- **The size and type of beaches:** beaches absorb some of the waves' energy and protect the coastline. Pebble beaches dissipate energy from waves through friction and percolation.
- **Human development:** sand and pebble extraction from beaches for use as building materials weakens a coast's protection and can lead to erosion. Sea walls, groynes and other coastal protection schemes may help to protect an area from erosion, but can also increase erosion further along the coast.

Knowledge check 20

What are the key differences between a flow, a slide and a slump?

Developments on the top of cliffs can increase runoff and cause instability and cliff failure.

Material is **transported** along a coastline by:

- **Traction:** large stones and boulders are rolled and slid along the seabed and beach by moving water.
- **Suspension:** very small particles of sand and silt, picked up by turbulence, are carried along by the moving water.
- **Swash and backwash**: sand and shingle are moved up the beach by swash and back down the beach by backwash.
- **Longshore (littoral) drift**: material is moved along the shoreline by waves that approach the shore at an angle. Swash moves sand and shingle up the beach at an angle but the backwash is at right angles to the beach. This results in material zigzagging its way along the beach according to the prevailing wave direction. Obstacles such as groynes interfere with this drift, and accumulation of sediment occurs on their windward sides, leading to entrapment of beach material. Deposition of this material also takes place in sheltered locations, such as where the coastline changes direction abruptly, creating sand spits.

Deposition occurs in low-energy environments, such as bays and estuaries. When sand is deposited on a beach – i.e. when the swash is more dominant than the backwash – and dries out, it can be blown by the wind further inland to form sand dunes at the back of the beach. In a river estuary, mud and silt can build up in sheltered water to create a salt marsh. Here the fresh water of the river meets the salt water of the sea, causing **flocculation** to occur and creating extensive areas of mudflats.

Coastal landscape development

Erosional landscapes

Cliffs, **headlands** and **bays** form when rocks of differing hardness are exposed together at a coastline. Tougher, more resistant rocks (such as granite and limestones) tend to form headlands with cliffs. Weaker rocks (such as clays and shales) are eroded to form sandy bays.

On an indented coastline, headlands and the offshore topography concentrate wave attack on that headland. Many headlands have a **wave-cut platform** between high and low tide, which can cause friction for the wave. However, due to headlands' solid nature they do not absorb energy, as a sandy beach would do, so waves can break at the foot of the cliff, causing maximum erosion. In a bay, waves have to travel further, and a beach absorbs wave energy and reduces the power of the wave before it reaches the cliff. Where there is a wide, deep, sandy beach, waves may not even reach the back shore at all.

A cliff base will be eroded faster than the rock above by the breaking waves in the tidal zone. A **wave-cut notch** develops here and rocks above may overhang. Gravity causes these overlying rocks to fall and the cliff line retreats, leaving the remnants of the rocks of the original cliff line as an almost macro-scale flat (less than 4°) base (a wave-cut platform). However, at a micro scale, most wave-cut platforms have a series of small terraces, micro-cliffs and deep rock pools.

Flocculation The process by which a river's load of clays and silts carried in suspension is deposited more easily on its meeting with sodium chloride in seawater.

Exam tip

Questions on this area of study will require an understanding of the distinctive nature of the processes in this environment. References to sequence and time are key here.

Exam tip

You are required to know examples of all of these landforms from the UK and other areas beyond the UK.

At any point on a cliff coastline where there is a weakness, such as cracks, joints or along bedding planes, erosion can take place. Where waves open up a prolonged joint in the cliff face, they form a deep and steep-sided inlet, or geo. If this inlet is widened to become a small **cave**, a crack at the back of the cave may open up like a chimney to the surface, creating a hole at the top of the cliff above. This is often referred to as a **blow hole**.

Where caves are created on headlands and are eroded back into the headland, they sometimes meet a cave that is being similarly eroded on the other side of the headland. The back wall separating the caves weakens and eventually the sea pushes straight through, forming an **arch**. The sea is now able to splash under the arch, further weakening it until eventually the arch roof collapses, leaving the seaward side of it as a separate island called a **stack**. Over time the stack will be eroded to form a **stump**.

Depositional landscapes

Coastal deposition takes place on sheltered stretches of coast. Sediment that can no longer be transported along the seabed or is suspended in water is deposited to form features such as **beaches** and **spits** which build up by **accretion**.

Beaches are commonly found in bays. Here, wave refraction creates a low-energy environment that leads to deposition. Beaches are made of either sand, shingle or a mixture of both. The characteristics of the available sediment and the power of the waves influence the nature of the deposition. Beaches are often classified as being either swash-aligned – where sediment is taken up and down the beach with little sideways transfer – or drift aligned, where sediment is transferred along a beach by longshore drift.

Beaches can be subdivided into different zones indicating the position of aspects of the beach in relation to breaking waves and tidal ranges:
- **offshore**: beyond the influence of breaking waves
- **nearshore**: intertidal and within the breaker zone
- **foreshore**: the surf zone
- **backshore**: usually above the influence of normal wave patterns, marked at the lower end by berms, and may have a storm beach further up

Within the offshore and nearshore zones, a number of minor beach landforms can be found:
- **Ridges and runnels** (alternate raised sand bar and dip sections) run parallel to the shoreline and are caused by periods of strong backwash. They are often exposed at low tide, but are hidden at high tide.
- **Beach cusps** are a series of small depressions that develop where beaches are made of sand and shingle. The sand is worn away more easily than the shingle and creates a semicircular hollow in a miniature bay-like formation. Once created they self-perpetuate, especially on swash-aligned beaches.
- **Ripples** are micro beach ridges parallel to the shoreline created by wave action in the foreshore zone on low gradient beaches.

Knowledge check 22

Explain why wave-cut platforms tend to have a maximum width of about 0.5 km.

Exam tip

A good way to demonstrate you know what each of these landforms looks like is to draw a sketch of it. Give it a go!

Accretion The growth of a natural feature by enlargement. In the case of coasts, sand spits grow by accretion as do other landforms such as sand dunes.

Exam tip

You will be asked to describe landforms. Refer to size, shape, nature of sediments and field relationship (i.e. where the landform lies in relation to the landscape).

Spits are long narrow stretches of sand or shingle which protrude into the sea. They result from materials being moved along the shore by longshore drift. This movement continues in the same direction when the coastline curves or where there is an estuary with a strong current which interrupts the movement of materials and they project out into it. The end of the spit is often curved (creating a series of laterals) where waves are refracted around the end of the spit into the more sheltered water behind. This is also helped by the direction of the second most dominant wind.

If a spit joins the mainland at one end to an island at the other it is called a **tombolo**. Where a spit has developed right across a bay because there are no strong currents to disturb the process, it creates a **bar** which then dams the water behind it forming a lagoon. Bars also develop as a result of storms raking up pebbles, and this shingle left in ridges offshore creates a type of **barrier beach**.

Offshore bars are deposits of sand and shingle situated some distance from a coastline. They usually lie below the level of the sea, only appearing above the level of the water at low tide. There are two explanations as to where and how they form:

- in shallow seas where the waves break some distance from the shore
- where steep waves break on a beach, creating a strong backwash that carries material back down the beach forming a ridge

When an offshore bar appears above the level of the sea for most of the time, it becomes a barrier beach. In this case, a series of elongated islands stands above sea level with a lagoon on the landward side and the ocean on the other.

Sand is often deposited by the sea under normal low-energy conditions. Wind may then move the sand to build dunes further up the beach, which in turn become colonised by stabilising plants – a process known as a **psammosere**.

Estuarine environments

Sheltered river estuaries, or the zones in the lee of spits, are areas where there are extensive accumulations of silt and mud, aided by the process of flocculation and gentle tides. These are **mudflats**, and such intertidal areas will be colonised by vegetation, and as with sand dunes, a succession of plant types may develop over time, called a **halosere**. The resultant landform is called a **salt marsh**. (Figure 8)

The initial plants of a halosere must be tolerant of both salt and regular inundation at high tide. These are called halophytes and include eelgrass and spartina grass. The latter has both a long root system and a mat of surface roots to hold the mud in place. These plants trap more mud and build up a soil for the next stage so that plants such as cord grass, sea lavender and sea aster can grow. As the mat of vegetation becomes more dense, the impact of the tidal currents reduces, humus levels increase, allowing reeds and rushes to grow next, and then alder and willow.

Salt marshes often have complex systems of waterways, known as creeks. In some extensive salt marsh areas, hollows of trapped seawater form, which then evaporates creating salt pans.

Psammosere The succession of plants that develops on a sand dune complex. Plants include sea rocket and lyme grass nearer the sea, with marram grass, fescue and gorse inland.

Exam tips

You will be asked to describe landforms. Refer to size, shape, nature of sediments and field relationship (i.e. where the landform lies in relation to the landscape).

A good way to demonstrate you know what each of these landforms looks like is to draw a sketch of it. Give it a go!

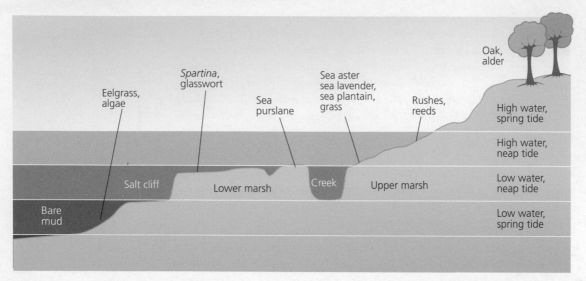

Figure 8 The structure of a salt marsh

Landscapes of sea-level change

Changes in sea level take place over time due to sea temperatures being colder or warmer than the present, or relative changes in land levels.

Eustatic change may result from a fall in sea level due to a new glacial period, when water is held as ice. This explains why during previous glacial periods the English Channel was dry land. At the end of a glacial period, the ice on land melts and global sea levels rise again.

Isostatic change arises from changes in the relationship of land to sea. During a glacial period, as ice collects on the land, the extra weight presses down on the land causing it to sink, and the sea level to rise. As the land ice melts the land begins to move back up to its original position (isostatic readjustment), and the sea level falls. This process is variable depending on the thickness of the original ice and the speed of its melting.

Tectonic processes associated with plate movement have also caused changes to sea level. By their nature, they tend to be quite localised. For example, the earthquake that struck Anchorage (Alaska) in 1964 increased the height of land by 9 m and lowered sea levels by up to 3 m in some areas.

In the last 10,000 years, a geological period called the Holocene – and especially between 10,000 and 6,000 years ago – saw the global sea level rise very quickly. It flooded the North Sea and English Channel, broke the link between Britain and Ireland and flooded many river valleys to give the distinctive indented coastline of southwest England and Ireland, known as rias (see below). Since then, sea levels have remained largely consistent.

The effect of a rise in sea level is to flood the coast, creating a coastline of **submergence.** Rising sea levels have flooded pre-existing valleys. **Rias** are drowned river valleys with long fingers of water stretching a long way inland, including the

Making connections

The concept of climate change is also discussed within the chapter on the carbon cycle (page 18).

Knowledge check 23

Describe the impact of isostatic sea-level change on the British Isles.

tributary valleys. They are widest and deepest nearer to the sea and get progressively narrower and shallower inland. They are often winding, following the shape of the pre-existing river valley. (Figure 9(a-c))

Fjords are glaciated valleys that were drowned by the rising sea level during the Holocene. They often have a shallower area at the mouth called a rock threshold, where the ice thinned as it reached the sea and hence lost its erosional power. Fjords have the typical steep-sided and deep cross profile associated with glacial troughs, and they can stretch many kilometres inland. The main channel is often straight, with right-angled tributary valleys. (Figure 9(d-f))

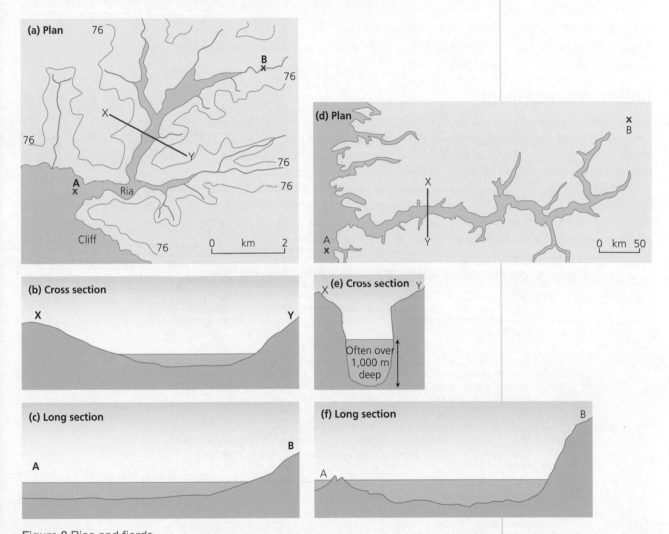

Figure 9 Rias and fjords

Dalmatian coast is the name given to a drowned coastline where the main relief trends run parallel with the line of the coast. The ridges of upland produce elongated islands separated from the mainland by the flooded valley areas. The name originates from the Adriatic coast of Dalmatia (Croatia).

The effect of falling sea levels is to expose land normally covered by the sea, creating a coastline of **emergence**. Cliffs that are no longer being eroded become isolated from the sea, forming **relic cliffs**. 'Fossil' features such as former caves and stacks are left higher up from the coast on raised **marine platforms** compared with present-day features. Less dramatic residual features are **raised beaches**. These are common on the coast of western Scotland, where a series of raised sandy and pebble-ridden terraces can be found above the level of current sea levels.

Sea-level rise associated with climate change is an important issue for the future, with increases of several centimetres (or more) expected over the coming decades. This sea-level rise is due to:

- the thermal expansion of water as it becomes warmer
- more water being added to the oceans following the melting of freshwater glaciers and ice sheets, such as those in Greenland

Coastal management

Some geographers classify coastal defence strategies against risks of coastal flood and erosion into '**hard engineering**' methods and '**soft engineering**' methods. Hard engineering approaches (Table 2) have significant financial implications and severe environmental costs, some of which are unpredictable. They tend to be focused on areas of greatest need, particularly urban areas with high land values and population densities. Soft engineering (Table 3) is less costly, and less environmentally damaging. For these reasons, it is often stated to be more sustainable. Many modern coastal management schemes combine elements that are both 'hard' and 'soft' in an integrated approach.

Table 2 Hard engineering strategies

Strategy	Description	Commentary
Sea wall	A concrete or rock wall at the foot of a cliff or at the top of a beach. It usually has a curved face to reflect waves back out to sea.	Although effective at the location where they are built, they deflect erosion further along the coast. They are expensive and have high maintenance costs.
Groynes	Timber or rock structures built at right angles to the coastline. They trap sediment being moved along the coast by longshore drift.	The beach created increases tourist potential, and gives protection to the land behind. However, this same process starves beaches further down the coast of sand and increases erosion there.
Rip-rap (rock armour)	Large, hard rocks dumped at the base of a cliff, or at the top of a beach. It forms a permeable barrier to breaking sea waves.	Relatively cheap, and easy to construct and maintain. The rocks used are often brought in from other areas and hence may not blend in.
Revetments	Wooden barriers, in a slat-like form, placed at the base of a cliff or top of a beach.	They are intrusive and very unnatural.
Gabions	Wire cages filled with small rocks that are built up to make walls. Often used to support small, weak cliffs.	They are relatively inexpensive. They look unsightly to begin with but as vegetation grows they blend in better. The metal cages rust and break easily.

> **Knowledge check 24**
>
> Make three lists: coastal landforms created by erosion; coastal landforms created by deposition; and coastal landforms resulting from sea-level change.

Exam tip

You will often be asked to 'evaluate' the effectiveness of management strategies. To do this, consider their advantages and disadvantages.

Table 3 Soft engineering strategies

Strategy	Description	Commentary
Beach nourishment	The addition of sand or pebbles to an existing beach to make it higher or wider. The materials are usually dredged from the nearby seabed and spread, or 'sprayed', onto the beach.	A relatively cheap and easy process, and the materials used blend into the natural beach. However, it is a constant requirement as natural processes continue to do their work.
Dune regeneration	The planting of marram grass and other plants that bind the sand together. Areas are often fenced off to keep people off newly planted dunes.	Maintains the look of a natural coastline, and provides important habitats. The process requires a lot of time to be effective.
Marsh creation	Low-lying coastal lands are allowed to be flooded by the sea. The area becomes a salt marsh.	It provides an effective buffer to the power of waves, creating a natural defence and an opportunity for wildlife habitats. However, agricultural land is lost and landowners require compensation.
Mangrove planting	In some tropical coastal areas, e.g. Indonesia and Queensland (Australia), mangroves have been planted to protect coasts from large swell waves, storm surges and tsunamis. Their extensive roots build up sediment and increase wildlife.	The power of waves is rapidly reduced as they pass through the vegetation, lessening damage. Belts, ideally thousands of metres across, can significantly reduce flood extent in low-lying areas. They reduce loss of life and damage to property when tsunamis occur.

Shoreline management plans

Shoreline Management Plans (SMPs) are an approach to coastal management that involves all stakeholders in making decisions about how coastal erosion and coastal flood risk should be managed. They aim to balance economic, social and environmental needs and pressures at the coast. There are 22 SMPs in England and Wales. In Scotland and Northern Ireland, the devolved governments and local authorities are jointly responsible for coastal protection.

SMPs provide a large-scale assessment of the risks associated with coastal processes and present a policy framework to reduce these risks to people and the developed, historic and natural environment, in a sustainable way. They predict, so far as it is possible, the way in which the coastline will be shaped into the future, defined as 100 years.

During the planning processes of SMPs, four coastal defence policies are considered:
- **Hold the line:** maintaining or upgrading the level of protection provided by defences.
- **Advance the line:** building new defence seaward of the existing defence line.
- **Managed realignment (or retreat):** allowing retreat of the shoreline, with management to control or limit movement.
- **No active intervention (sometimes referred to as 'do nothing'):** a decision not to invest in providing or maintaining defences.

SMPs are an example of integrated coastal zone management (ICZM). ICZM requires:
- that an entire coastal zone is managed, not just the narrow zone where waves cause erosion or flooding, including all ecosystems and human activity in the zone
- recognition of the importance of the coastal zone to people's livelihoods as large numbers of people live and work at the coast
- recognition that management of a coastal zone must be sustainable – economic development has to take place to improve the quality of life of people and this needs to be environmentally appropriate and equitable

Exam tip

For many of the strategies of coastal management identified, have an example of where they have been employed.

Exam tip

Note that the SMPs are closely linked to natural sediment cells (see page 34).

Knowledge check 25

Who manages Shoreline Management Plans (SMPs) and why is this important?

Case studies

You are required to consider two case studies:

1 A case study of a coastal environment at a local scale to illustrate and analyse fundamental coastal processes and the landscapes that have been created by these processes. It is important that you engage with field data within this location. You should also consider how such areas can be managed in a sustainable way, as such locations are often popular with tourists and yet used by locals. Examples in the UK include the Yorkshire coast such as Flamborough Head, or the Dorset Jurassic Coast.

2 A case study of a coastal environment beyond the UK to illustrate and analyse coasts as presenting risks and opportunities for human occupation and economic development. It is important that you consider how the people of this area have responded to the risks and opportunities, within the key themes of resilience, mitigation and adaptation. Examples could include the delta coastline of the Netherlands, the Sundarbans (Bangladesh) or the coral reefs off northern Australia.

Exam tip

The exam questions on this case study are likely to use one or more of the following words: sustainable, resilience, mitigation and adaptation. Make sure you understand these terms.

Summary

After studying this topic, you should be able to:

- explain how coasts operate as natural systems, and be able to describe their main inputs (wind, waves and tides) and outputs (landforms)
- discuss the range of geomorphological processes that operate in coastal environments, including weathering, mass movement, erosion, transportation and deposition by the actions of wind and waves
- describe the variety of landscapes that have been created in coastal environments over time, largely due to the erosional and depositional actions of wind and waves
- explain the variety of mechanisms of sea-level change (eustatic, isostatic and climate change) and the amended landforms due to submergence and emergence
- discuss the concept of coastal management against flood and erosion, and the variety of approaches by which it can be addressed sustainably

▇ Glacial systems and landscapes

Glaciers as natural systems

Glaciers and their immediate fringes create very distinctive **landscapes**, many of which remain visually stunning for thousands of years after their creation. Within the label of glacial, we also consider the landscapes created by meltwater from glaciers and ice sheets (**fluvioglacial**) and those created by very cold climates, even though glaciers may not have been involved in their creation (**periglacial**).

The most obvious **input** into the glacial system is snow. Other forms of precipitation can also fall on a glacier, adding to the mass when they freeze. This overall gain in mass is called **accumulation**. A glacier also has rock fragments on its surface and embedded within and under the ice. This is also part of the mass of a glacier and is important, as it is the tool that carries out erosion. Snow and rock can fall onto the surface of a glacier as a result of avalanches or from weathering processes like frost shattering on a valley's sides. Solar energy is also an input into the system. Ice and snow have a very high **albedo**, but the sun's energy still leads to melting and evaporation, and the loss of mass.

Exam tip

Once you have completed this option, be clear of the difference between glaciation, fluvioglaciation and periglaciation.

A glacier's main **outputs** are through melting, sublimation and evaporation. Sublimation occurs when ice is converted to water vapour without melting. This can take place when it is cold and the atmospheric conditions are dry. Most of this loss is very seasonal occurring during the summer. The collective term for this output is **ablation**. Material can also be transported away in meltwater streams. These are often a brown colour, reflecting their high load. Less obvious are the milky blue streams flowing from glaciers. They have this coloration due to the high content of very finely ground-down rock material called **rock flour**. Energy is also transferred through reflection from snow and ice, and also radiation.

The main **store** of a glacial system is the glacier itself, though it will also include meltwater contained above, within and below it, as well as weathered and eroded material that forms part of the glacier's mass.

Transfers involve the movement of ice and water in the glacial environment. They are best considered under the term 'glacial budget' (see page 47).

The nature and distribution of cold environments
The global distribution of cold environments, present and past

Cold environments are typically those of the high latitudes, caused by their proximity to the North and South Poles. The equatorward boundary of cold climates is generally taken as the line where the mean temperature of the warmest month is no more

Exam tip

Note that this topic should be illustrated with examples from both the UK and beyond the UK.

Landscape An expanse of land/scenery that can be seen in a single view. It covers all aspects of the view – both natural landforms and human-created features.

Albedo The amount of incoming solar radiation that is reflected by the Earth's surface and atmosphere. Fresh snow and ice have an albedo as high as 90%.

Making connections

It is important you apply the principles of systems theory (page 6), including feedback and dynamic equilibrium, to this environment.

Making connections

Processes within the water cycle were explained on page 10. Although the glacial context is different, note that several of the processes are applicable in this environment.

than 10°C. This line broadly corresponds with the poleward limit of tree growth. In the northern hemisphere this isotherm swings south of the Arctic Circle into Alaska, Labrador, Greenland, Iceland and parts of northern Russia. In the southern hemisphere the only area of extensive cold is Antarctica. Other cold environments exist in the mountainous areas of the world – the Andes, Rockies, Alps, Himalayan mountains and Tibetan plateau.

To many people cold environments are synonymous with ice caps, sheets and glaciers. Around 10% of the world's surface is covered by ice but over the past few million years the extent of such ice has been much larger than now, covering about one-third of the Earth's surface.

Glaciers are important because they cover substantial parts of the Earth's surface and because they have left their imprint on the land they formerly covered.

During the last major ice advance, 21,000 years ago, much of the northern hemisphere was covered by ice and glaciers. These covered nearly all of Canada and much of northern Asia and Europe, and extended well into the USA – in total 30% of the land surface.

Current glaciated areas include the ice cap of Antarctica and northern Greenland, northern Russia, northern Scandinavia, northern Canada, and the high mountain areas of the high Andes, Rockies, Himalayas and Alps. Past glacial and periglacial environments include Great Britain and most of central Western Europe, where lower latitudes and altitudes show the imprint of the last (**Pleistocene**) glaciations.

Physical characteristics of cold environments

Cold environments can be subdivided into two types: ice caps and tundra.

Ice cap environments

Here average temperatures in all months are below freezing (0°C) so that vegetation growth is impossible, and a permanent ice and snow cover prevails. At the South Pole the warmest month (December) has a temperature of –28°C and the three coldest months (July, August and September) an average temperature of –59°C. Thermometers have been known to record temperatures of –90°C. Precipitation is scant in amount as the low temperature, low humidity levels and the extreme stability of the air all inhibit snowfall. Precipitation levels are recorded as less than 150 mm per year.

Tundra

This is the name given to a climatic and vegetation type that can be found in the most northerly parts of North America and Eurasia (north of 65°N). The main features of the climate are:

- long and bitterly cold winters with temperatures averaging –20°C
- brief, mild summers with temperatures rarely being above 5°C
- a large temperature range of over 20°C
- low amounts of precipitation, less than 300 mm, most of which falls as snow
- strong winds blowing the dry powdery snow in blizzards, and creating a high **wind-chill** factor

Knowledge check 26

In the context of glaciers, explain the difference between accumulation and ablation.

Knowledge check 27

Distinguish between the following types of cold environments: glacial, periglacial and alpine.

Pleistocene A geological time period stretching from 2 million years BP (before present) to 10,000 years BP. It was characterised by a series of alternating cold phases (glacials) and warm phases (interglacials), collectively known as the ice ages.

Wind-chill The accentuation of cold temperatures when accompanied by high wind speeds. For example, an air temperature of 0°C with a wind speed of 6 ms⁻¹ has a chill factor of –10°C.

Although the winters are severe and the sea regularly freezes, temperatures do not fall as low as areas further inland due to the moderating effect of the sea. The cold temperatures are due to the short hours of winter daylight in such areas, and although the daylight hours are longer in summer, the angle of inclination of the sun is low. High pressure with dry descending air dominates these areas. In summer, some depressions do penetrate, giving some precipitation.

Tundra areas are associated with large areas of frozen soil – **permafrost** (see periglaciation – page 53). Due to the climate and the soils, they have a distinctive vegetation type which features:

- dwarf species such as cotton grass, mosses and lichens growing close to the ground, often in a cushion-like form – an adaptation to the strong winds that blow in the area
- dwarf willows and stunted birch trees in sheltered places
- plants that have long dormant periods during the cold dark winters, but grow rapidly in the summer when daylight hours are long. 'Bloom mats' of anemones, Arctic poppies and saxifrages burst into life, providing a mass of colour
- plants with small leaves to limit transpiration
- plants that are shallow rooted because the soil is permanently frozen at a shallow depth

Systems and processes
Glacial budgets

The **glacial budget** is the relationship between accumulation and ablation:

- If accumulation is greater than ablation, the glacier gains mass and the glacier snout advances.
- If ablation is greater than accumulation, the glacier loses mass and the glacier snout retreats.
- If accumulation is equal to ablation, the snout of the glacier remains in the same place – it is stationary.
- An **equilibrium line** can be shown on a glacier where at any one time the amount of accumulation equals that of ablation. This is also known as the **firn line**.
- Hence there are spatial variations in a glacial budget. Most of the accumulation occurs at the upper end of a glacier. At higher altitudes, where it is much colder this exceeds losses due to ablation. There is therefore **net accumulation** and this part of a glacier can be called the **zone of accumulation**. At the lower end of a glacier, the reverse is the case: ablation at warmer, lower altitudes exceeds accumulation, so there is **net ablation** and this part of a glacier is called the **zone of ablation**.

> **Exam tip**
>
> It is important that you understand fully the interplay between accumulation and ablation, including the seasonal variations in each.

Note that there are usually seasonal changes in the glacial budget. Glaciers usually have a positive balance in winter, when more snow accumulates, and a negative balance in summer, when there is more ablation due to higher temperatures.

> **Exam tip**
>
> Note that glaciers do *not* retreat – it is their snout that retreats. Ice continues to move downhill but melts at a faster rate.

Historical patterns of ice advance and retreat can be examined for any one of the world's glaciers, and each glacier can have a different pattern of advance and retreat.

Types of glacier

The largest glaciers are called **ice sheets**. These have a flattened dome-like cross section, are hundreds of kilometres in width and cover an area of more than 50,000 km². The most well-known ice sheets, because of their size, are those of Antarctica and Greenland. During the Pleistocene there were also two other enormous ice sheets, one over Scandinavia and Great Britain, and another over much of North America. Smaller domes, less than 50,000 km², are called **ice caps**. An **ice shelf** is an area of thick floating ice sheet attached to a coastline, whereas an **ice field** is a relatively flat and extensive mass of ice.

The other main type of glacier is the one that occupies a valley or lowland basin. These can further be divided into the following:

- **Valley glacier:** body of ice that moves down a valley under the influence of gravity and is bounded by rock walls on either side.
- **Cirque glacier:** small body of ice that occupies an armchair-shaped hollow in mountains, which has been cut into bedrock.
- **Diffluent glacier:** valley glacier that diverges from a main glacier and crosses a drainage divide through a diffluence col.
- **Piedmont glacier:** a glacier that leaves confining rock walls and then spreads out to form an expanded glacier at the foot of a mountain valley.

An alternative classification of glaciers is according to their temperature:

- **Warm-based/temperate/alpine glaciers:** found in areas where temperatures are high enough to cause some summer melting. This allows the glaciers to advance in winter and retreat in summer. The meltwater reduces friction and lubricates the base, allowing the glacier to move more quickly. The associated increase in energy allows more material to be transported and therefore some erosion and deposition occurs with these types of glacier.
- **Cold-based/polar glaciers:** found in areas where the temperature rarely rises above freezing so there is little melting or movement and much less erosion transport and deposition.

Geomorphological processes

Glacier formation

When snow crystals fall they have an open, feather-like appearance and a low density. However, if the snow crystals are compacted by the weight of overlying snow, or if they are partially melted, they are converted to a mass of partially consolidated ice crystals with interconnected air spaces between them. Such material is called **firn** or **neve**. As this process develops further it becomes more dense, until most of the air spaces are eliminated and pure ice develops.

Most glaciers are composed of the ice produced from snow that has been modified in this way, but on some glaciers refrozen meltwater can make up much of the mass.

Knowledge check 28

Compare the typical dimensions of valley glaciers and cirque glaciers.

Eventually sufficient ice may accumulate so that, under the influence of gravity, the mass begins to move and thereby becomes a glacier.

Glacial movement

Glaciers are generally not static. Most glaciers move slowly – less than 50 m a year. However, some glaciers do surge and may for short periods of time reach speeds of 5 m per hour, sometimes more.

The movement of a glacier takes place in three main ways:

■ **Basal sliding** is linked to the pressure melting of ice that takes place along its base. Ice normally forms from water at a temperature of 0°C, but the temperature at which water freezes is reduced under pressure. As a glacier moves, it will exert pressure and therefore some melting may take place at its base. A thin film of water may then exist between the glacier and the bedrock. This film reduces friction – acts as a lubricant – and allows the glacier to slide. Such movement is more likely to occur in warm-based glaciers, where temperatures are close to their melting point. Linked to this process is **regelation**. This is the refreezing of water under a glacier when pressure is diminished in the lee of an obstruction. Pressure melting takes place on the upstream side, creating a film of water, and regelation takes place on the downstream side, and the combination of both processes allows a temperate glacier to slide downhill.

■ **Compressional and extensional flow** take place when ice cannot deform sufficiently quickly due to stresses beneath the ice. For example, there may be rapid changes in the gradient of the bedrock – steep sections followed by more gentle sections. As a result, ice fractures and movement take place along a series of planes. Tensional fractures (caused by extension) create crevasses in the upper layers of the glacier. Shear fractures (caused by compression) result in the thrusting up of ice blocks along faults. The outcome of this form of flow is a very irregular surface to the glacier.

■ **Internal deformation** is much smaller in scale and takes place in all glaciers, but it is the most common form of movement in cold-based glaciers. It takes place where the ice crystals set themselves in line with the movement of the glacier and slide past each other. Movement can occur along lines of weakness called cleavage planes. This is sometimes called **laminar flow**, where it can be linked to layers of annual accumulation.

Glacial weathering and erosion

A glacier performs erosion as it has both the materials and the energy to do so. A glacier receives material through a combination of weathering and rock fall. Weathered material comes from frost action and from nivation:

■ **Frost action** occurs where diurnal temperatures hover around freezing point (above during the day and below at night). Water in cracks in the rocks freezes at night, widening the crack by expansion (usually by 9%). In the day, more water collects in the crack, which again freezes, widening it further. The rock then shatters into angular fragments that fall onto the ice and are used as the tools for erosion by abrasion, or they form mounds at the base of the slope, forming **scree**. At a much smaller scale, when water freezes in a rock, the ice attracts very small

Knowledge check 29

Describe the variations of movement within a glacier both in plan view and with depth.

Knowledge check 30

Glaciers can also move by surges. Describe and explain such movements of ice.

particles of water that have not frozen from the adjoining pores and capillaries. Nuclei of ice crystal growth are thus established, and some believe this form of ice crystal growth to be a more potent form of rock shattering.

■ **Nivation** is where physical and chemical weathering under a patch of snow, due to diurnal and seasonal temperature changes, causes the rock to disintegrate. Subsequent meltwater washes away the weathered material, and a small hollow, called a nivation hollow, deepens.

Exam tip

Although weathering and erosion operate together, in an exam context you need to be clear of the differences between the two sets of processes.

Although glacier ice itself does not cause marked erosion of a rock surface, when it carries the debris that has either fallen onto it, or washed into it, abrasion occurs. Sharp rock fragments carried by the ice embed themselves in the base and sides of the ice. This is used to grind down the bedrock like sandpaper, making it smooth. It can also leave scratches on the rock in the direction of ice movement, called **striations**. Much of the debris in glaciers is ground down to a fine mixture of silts and clays, known as rock flour.

Glaciers can also cause erosion by **plucking**. If the bedrock beneath the glacier has been weathered in periglacial times, or if the rock is full of joints (well-jointed), the glacier can detach large particles of rock and take them with it. As this process continues some of the underlying joints in the rock may open up still more as the glacier removes the overburden of dense rock above them – a process known as **pressure release**.

Glaciated landscape development
Landforms and landscapes due to glacial erosion

As debris-laden ice grinds and plucks away the surface over which it moves, characteristic landforms are produced, which give distinctive glacial landscapes. One of the most impressive landforms is the **cirque** (**corrie**, **cwm**). This is an armchair-shaped, steep-sided hollow at the head of a glaciated valley. They are often N or NE facing in the northern hemisphere, as this direction will receive the least sunlight, and are in the lee of prevailing winds, causing the snow to accumulate longer. Their size varies but they are often around 500m in diameter with a back wall up to 1,000m in height.

The hollow is deepened initially by nivation and then, as the ice accumulates, by rotational slide (Figure 10). This enhances the abrasion process due to increased compression in the base of the hollow. Ice pulls away from the back wall, plucking rocks already loosened by freeze–thaw weathering. This creates a **bergschrund** – a crevasse that forms where the ice starts to pull away from the back wall of the cirque. Material falls down this crack and is embedded in the base of the ice, which is then used to abrade and help deepen the hollow. Where the pressure and erosion are lower at the front edge of the hollow there can be deposition of **moraine** at the lip of the cirque, which often allows water to accumulate behind it in a post-glacial period, forming a small circular lake (**tarn**).

Moraine The collective term for the debris which a glacier transports and then deposits.

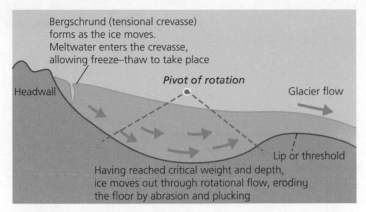

Bergschrund (tensional crevasse) forms as the ice moves. Meltwater enters the crevasse, allowing freeze–thaw to take place

Pivot of rotation

Headwall

Glacier flow

Lip or threshold

Having reached critical weight and depth, ice moves out through rotational flow, eroding the floor by abrasion and plucking

Figure 10 Formation of a cirque (corrie) by rotational movement

As cirques develop they eat back into the mountain mass in which they have developed. When several cirques lie close to one another, the divide separating them may become progressively narrowed until it is reduced to a narrow precipitous ridge called an **arête.** Should the glaciers continue to erode away at the mountain from all sides, the result is the formation of a **pyramidal peak** or **horn**.

Glacial valleys (or **troughs**) develop where glaciers flow into pre-existing river valleys. They widen and deepen the original valley, making it steep-sided with a wide, flat base. Glaciers tend to straighten the valley, cutting off spurs and leaving cliffs as **truncated spurs**. At the upper end of the valley, where the glacier has entered the valley from the corries above, there is often a steep wall called the **trough end**. Glacial flow is uneven and where compressing flow has over-deepened a section of the valley, or where there is softer rock which is more easily eroded, it forms a rock basin, which often fills with water post-glacially forming a **ribbon lake.** A **hanging valley** forms where ice in a tributary valley cannot erode effectively because its movement is blocked by ice in the main valley (or it may be smaller in size). It therefore remains 'higher' and smaller than the main glacial trough (hanging above it). Post-glacially, a hanging valley may be identified from a waterfall into the main valley below.

Roche moutonnées are isolated rocks, generally 5–30 m in height, along the base of a glaciated valley. They are characteristically smooth on one side and jagged on another. They form where a glacier moves over a band of harder rock: it smooths the upstream side by abrasion, often leaving **striations** to show the direction of movement. As the glacier spills over the top of the obstruction it removes the loose rocks by plucking, leaving a jagged edge on the downstream side.

A number of high-latitude coastlines, such as those in Norway, western Scotland and South Island, New Zealand, are flanked by narrow troughs called **fjords**. These are submerged glacial valleys. Some of them are extremely deep (over 1,000 m in depth) and bear witness to the substantial erosional powers of glaciers. These great depths are usually separated from the open oceans by a sill of solid rock, or **threshold**, where the depth is only 200 m. This is due to the erosional power of the glaciers becoming much less as they met the sea.

Exam tip

You are required to know examples of all of these landforms from the UK and other areas beyond the UK.

Exam tip

You will be asked to describe landforms. Refer to size, shape, nature of sediments and field relationship (i.e. where the landform lies in relation to the landscape).

Glacial transportation and deposition

The debris transported by ice may be divided into three main categories:

- **englacial debris**: material carried within a glacier
- **supraglacial debris**: material carried on a glacier surface
- **subglacial debris**: material carried at the base of a glacier

Deposition of this debris is a complex process. In general, deposition occurs when the carrying capacity of the ice decreases because the gradient decreases, the velocity of the glacier decreases, or more material becomes available than can be transported. The material deposited directly from the ice, which is smeared on to the landscape, is called **till** (or **ground moraine**). It consists of a wide range of grain sizes, from clays to stones, and is often called **boulder clay**. It also possesses very little stratification, frequently contains far-travelled erratic material and tends to have clasts (or stones) with edges and corners blunted by abrasion. Such till often has its larger particles showing an orientation, or alignment, to the movement of the glacier.

Deposited till may form a variety of distinctive landforms. Ridges of glacially deposited material occur along the end (terminal) margins of glaciers and ice sheets – these are collectively called **end moraines**. A series of end moraines may be traced across a lowland region for hundreds of kilometres, and may be up to 100 m in height.

Other moraines are associated with glaciers:

- **Lateral moraine:** moraine produced at the side of the glacier, having rolled there after falling on top of the ice or having fallen onto the side from frost shattering of the valley sides.
- **Medial moraine:** found at the confluence of two glaciers where two lateral moraines merge to form a deposit down the middle of the main glacier.

Drumlins

A **drumlin** is an oval-shaped hill, often of unsorted till, shaped like an egg half buried along its long axis. It can be up to 50 m in height and around 1,000 m long. Its long axis runs parallel to the direction of ice flow, with its steeper 'stoss' end pointing up-ice and its gently sloping 'lee' end pointing down-ice. Drumlins can be either rock-moulded or composed entirely of glacial or pre-existing sediments. They are often found in groups or swarms where the landscape is described as 'a basket of eggs' topography.

Geographers still do not know exactly how drumlins are formed beneath the ice. The most widely accepted theory suggests that different sediments, with different internal strengths, behave in different ways when overlying ice applies stress to them. This leads to zones of stronger sediments able to resist deformation, and these act as cores that develop into drumlins.

Erratics

Erratics are rocks which have been transported by a glacier or an ice sheet and deposited in an area of different geology to that of its source. Therefore, they can be used as evidence to indicate the direction of ice movement. Rocks from Ailsa Craig, in the Firth of Clyde, have been found in southwest Lancashire, England, 240 km from their source.

Fluvioglacial processes and landforms

Meltwater flowing subglacially is often at high pressure. It may erode spectacular subglacial channel networks. The pressure normally forces meltwater towards the snout of the glacier, so subglacial channels can be used to trace the direction of ice flow. Lateral meltwater channels form when streams flow along the sides of the glacier or ice sheet, often following the contours of the slope, perched on the valley sides in parallel series.

A number of landforms are found in a **fluvioglacial** landscape:

Eskers

These are the infillings of ice-walled subglacial, englacial or supraglacial river channels. When the ice melts away it leaves behind an elongate, sinuous ridge composed of sand and gravel, which can be up to 100 m in height and extend for tens of kilometres. Eskers may form single channels or branching networks. Eskers formed subglacially often have up-down long profiles that reflect the dominant influence of the ice/surface slope. Supraglacial and englacial eskers often have beaded (or flattened) shapes, reflecting the slow melt-out of underlying ice.

Kames

These are features produced at the margins of a glacier or ice sheet. They consist of irregular undulating mounds of bedded sands and gravels that are essentially a group of alluvial cones or deltas deposited unevenly along the edges of a decaying ice sheet.

Proglacial meltwater channels

These are eroded by streams draining away from the ice margin. They follow the slope of the land and are often wide and deep. They are also often associated with lakes dammed by ice, that then spill over.

Meltwaters can create landforms known as **outwash plains** (**sandurs**) and **ice contact deltas**. Sandurs are large areas of often sandy and pebbly material washed out of the ice by glacial meltwater streams. These streams carry a lot of debris in braided channels, which leave the largest deposits nearest the ice front and carry finer particles further across the plain. The deposits are also affected by seasonal melting and appear in layers (**varves**), showing the larger deposits occurring each spring with the increased meltwater. Such deposits are less angular than those left by the ice itself, and are sorted by the water leaving deposits over a wide area. During their creation, residual blocks of ice are often left behind by a collapsing glacier. These melt over time, leaving small, often circular lakes, known as **kettle lakes**.

Periglacial processes and landforms

Examples of **periglacial** environments include those in high mountain regions such as the Rocky Mountains of North America, the interior plateaux of central Asia, and the tundra of Canada, Alaska and Russia.

Periglacial environments are associated with **permafrost**. This is the permanently frozen subsoil lying beneath the surface of periglacial areas, which underlies approximately 20–25% of the Earth's surface. Permafrost can be continuous,

Fluvioglaciation The action of meltwater from glaciers and ice sheets.

Exam tip

It is important to appreciate the differing field relationships of these features: eskers are formed within a glacier, kames at the sides, and proglacial means 'beyond the glacier'.

Exam tip

Fluvioglacial deposits are generally sorted, often dominated by rounded or sub-angular pebbles, gravels and sands. The larger deposits are often orientated downstream.

Periglaciation The processes and landforms on the fringe of ice sheets and glaciers, as well as those areas where freezing conditions dominate.

discontinuous or sporadic, depending on the size of the area that it covers. The **active layer** lies above the permanently frozen area and temperatures in this layer rise above freezing in the summer. The active layer varies in thickness, but is usually between 5 m and 15 m in depth. One factor that can influence the depth of the permafrost is the amount of vegetation cover – it decreases with increasing levels of vegetation.

Several landforms are characteristic of periglacial areas. These are outlined below.

Pingos

These are domed mounds of layered sediments with a core of ice and are usually up to 100 m in diameter, though they can be as wide as 2 km. Pingos can be open or closed. **Open pingos** can be seen in the Canadian Arctic and Siberia and are more likely to be found in clusters. **Closed pingos** are less common and are found in Alaska and Greenland. Whether a pingo is open or closed depends on its formation. Open pingos are formed in areas of discontinuous permafrost when groundwater from small areas of **talik** moves upwards because of an increase in hydraulic pressure as the active layer refreezes in winter. As the water freezes near the surface, it causes the ground surface to dome upwards, forming the characteristic shape of a pingo. Closed pingos are formed in areas of continuous permafrost when water pushed down from frozen lake sediments accumulates at depth, freezes and then expands again causing the ground surface to dome.

Some pingos have a small depression at their top, resulting from localised melting there. Over time a pingo may collapse and fracture, leaving a ring-shaped scar in the landscape.

Patterned ground

This is one of the most common periglacial landforms, and can be found in Alaska, Greenland and northern Canada. Patterned ground can take the form of ice-wedge polygons or stone polygons.

Ice-wedge polygons

Ice-wedge polygons are generally 20–30 m across and are formed in areas of continuous permafrost by the effect of the ground freezing in the winter and thawing in the summer. When temperatures fall in winter, water in fissures in the active layer freezes and therefore expands, pushing the ground apart. The wedge shape is maintained as water further down from the surface, over 3 m down, is cooler and remains frozen. Indeed, during the spring and summer more water and fine sediment flow down to this zone to refreeze, thereby perpetuating the process. The annual repetition of this process causes the gradual formation of marked ice wedges, which develop in polygonal patterns.

Stone polygons and stripes

Stone polygons are smaller than ice-wedge polygons – less than 10 m across – and occur in both permafrost areas and high mountain areas. These are also the result of freezing and thawing, being generated by frozen soil and the development and expansion of small ice lenses under the surface (collectively known as **frost heave**):

■ During the winter soils freeze downwards from the surface, reaching and 'grabbing' individual stones, which are pulled upwards by the vertical expansion of the frozen soil above them. The empty space left beneath is filled with loose unfrozen soil, so the stone is prevented from moving back when the soil below freezes.

Active layer The upper layer of the soil in permafrost areas where there is seasonal thawing.

Knowledge check 32

Distinguish between continuous, discontinuous and sporadic permafrost.

Talik A localised area where the permafrost has thawed.

■ Stones are also pushed towards the surface due to the pressure of small ice lenses growing beneath them. This occurs because stones have a lower specific heat and become colder more quickly than fine-grained soil, so ice will form first directly beneath them pushing them upwards. Stones also warm up more quickly so in the summer thaw, the ice in the soil beneath the stone melts first, allowing wet sediment to slump to fill the space beneath it. This prevents it from sinking back into its original position.

Exam tip

Be clear about the differences between the various types of polygon. Ice-wedge polygons are found in stone-free soils such as former lake beds; stone polygons in stone-rich soils.

The net outcome of both of these processes is that stones move upwards in the talik. As the stones collect on the surface of the ground the larger stones are pushed towards the edge of the pile by ground expansion and gravity, and smaller stones, sands and silt are left in the middle, resulting in a polygonal shape. When polygons develop on a slope, the shapes become elongated and a sorted stone stripe is formed.

Solifluction lobes

In areas where there is discontinuous permafrost, the subsoil does not thaw in summer so surface meltwater cannot percolate into the rock beneath. The soil becomes saturated and starts to flow downhill on even the gentlest of slopes (as low as 1°) – a form of mass movement. Reduced friction between particles due to the water content and a lack of vegetation to hold the soil together allows the soil to move, leaving tongue-like protrusions called **solifluction lobes**. Where the slope is steeper on the valley sides, **solifluction terracettes** form.

Other periglacial landforms

In some areas, **blockfields** (**felsenmeer**) occur. Here the larger angular blocks of stone released by frost shattering on a cliff slowly spread across flat plains below, forming a field or sea of isolated rocks.

Thermokarst refers to irregular hummocky terrain often studded with small water-filled depressions created by the melting of ground ice. Its development is the result of the disruption of the permafrost and an increase in the depth of the active layer. The creation of thermokarst may increase substantially with climate change, and cause many buildings and features such as roads, bridges and pipelines to become much more unstable.

Human impacts on glaciated landscapes
The concept of environmental fragility

Fragile environments are those where any disruption to the ecosystem, however slight, can have serious short-term and longer-term effects. The tundra is considered to be a fragile environment because of its climate and limited biological productivity. The slow rate of plant growth means that any disruption to the natural equilibrium of the ecosystem can take a long time to be corrected, if at all. The low productivity of the area together with limited **biodiversity** mean that plants are very specialised and disruption causes difficulty when it comes to plant regeneration.

Making connections

The concept of climate change is also discussed within the chapter on the carbon cycle (page 18).

Knowledge check 33

Make four lists: glacial landforms created by erosion; glacial landforms created by deposition; fluvioglacial landforms; and periglacial landforms.

Biodiversity The variety of different forms of life within an ecosystem.

Traditional activities in cold environments

The traditional economic activity of the indigenous population of the tundra was hunting and fishing. In the north of North America, the main activity of the **Inuit** was hunting seals, which provided them with meat, oil and skins. Fishing (including whale hunting) was also a major activity. The number of Inuit was small in terms of the vast area in which they lived so very little pressure was placed on the environment, which remained relatively undisturbed.

In the north of Europe, the **Sami** people of northern Scandinavia followed the seasonal movements of the herds of wild reindeer that provided them with most of the food and materials that they needed. Fishing was used to supplement their diet. Reindeer spend the winter period in the boreal forests living on tree mosses, lichen and bark. They then moved back into the tundra during the summer, a migration often involving distances of 300–400 km. Like the Inuit, the Sami lived in an environment that provided a sustainable way of life but which could only support a low-density population.

However, in recent years their way of life has begun to change. Sami art objects and handicrafts are sold to tourists. In the winter pastures, lichen from old trees, an essential part of the reindeer diet, is in short supply due to modern forestry methods. Reindeer are kept in enclosures and fed hay and other fodder during the winter. It is no longer a nomadic life, and many Sami now live in settled areas.

Modern activities in cold environments

The Alps is an example of a high-altitude cold environment. There are often conflicts between tourists and farming, and even different tourist groups in this area. Skiing and snowboarding off piste can set off avalanches, and tourist hotels, cable cars and ski lifts are said to ruin the landscape. The development of mountain sports and tourism in these remote areas has improved the transport network and economy by developing jobs in the tourist industry, which has kept more people living in the area and prevented further rural depopulation. However, many people feel that the fragile environment has been damaged, with the loss of rare plants like orchids and birds such as the golden eagle.

Alaska has vast reserves of oil and gas. They were first discovered around Prudhoe Bay which was inaccessible to ships due to pack ice. A pipeline needed to be built to the ice-free port of Valdez on the south coast of Alaska. This was constructed on insulated legs to avoid the warmed oil (warmed, so that the oil does not freeze in such low temperatures) from melting the permafrost and causing the pipeline to sink and break, and was raised on stilts to avoid blocking migration routes of the caribou. It also zigzags to avoid rupturing by ground movement due to either frost heave or earthquakes.

Modern buildings are insulated or raised above the ground with piles driven into the permafrost to avoid the ground warming up and developing thermokarst. Utilidors (insulated boxes) are used, elevated above the ground, to carry water supplies, heating pipes and sewerage between buildings, and are designed not to melt the permafrost.

Exam tips

You should focus on the *differences* between traditional activities and modern activities.

Research recent political decisions in the Alaskan Arctic – for example, in 2020 President Trump allowed more oil exploration to take place here.

The impact of climate change

As climate change has occurred, scientists have measured a retreat in the permafrost zone. This is particularly acute in northwestern Canada and in Siberia. Increases in temperature of only 1°C have led to the trebling of the thaw rate in some parts. The immediate impacts of melting can be seen in Arctic communities – buildings become undermined, roads subside unevenly and crack, and supports holding pipelines shift and even crack the pipelines.

Another major concern regarding the melting of permafrost is the release of organic carbon. The soils of the permafrost are normally crammed with un-degraded, well-preserved organic matter in the form of leaves, twigs, roots etc. This is an enormous store of carbon, kept inert by being frozen in the ground. However, as the ground melts and the organic matter starts to rot, carbon is released as either carbon dioxide or methane, creating more greenhouse gases (GHGs). This will melt more permafrost and so on, in a worsening positive feedback cycle.

Exam tip

Revisit the section on feedback mechanisms in the earlier part of this book (page 6).

The Arctic is estimated to contain about 900 Gt of carbon. Humans emit about 9 Gt of carbon from fossil fuels and deforestation every year. Hence, it would only take the release of 1% of carbon in Arctic permafrost soils to effectively double emissions of GHGs, especially methane.

Management of cold environments

The Arctic Council provides an important regional forum for the indigenous peoples of the Arctic to advance their agendas, including matters relating to the use of resources. This international convention provides a basis for groups to go back to their home countries and demand the rights specified by it.

In Alaska, First Nation bodies own about 12% of the land, whereas the US federal government owns about 60%. The latter currently has a preference for subsistence use regarding resources. It enables indigenous rights to catch marine mammals, and has established community quotas to ensure that local groups have a stake in the region's marine fisheries. Oil and gas developments have provided some long-term jobs for local residents. Such extractive industries have also provided substantial cash amounts to local First Nation communities.

In Greenland, there is no system of private property or well-defined rights to land and natural resources, but the indigenous-controlled Greenland Home Rule body has authority to make most decisions about the use of terrestrial and marine resources.

In Scandinavia, national institutions have traditionally regulated natural resources, so native stakeholders have had greater difficulty establishing their rights. For example, an ongoing struggle to secure rights to land and natural resources by the Sami has met with limited success. This is especially the case in Sweden, where the courts have denied claims to indigenous rights despite state recognition of these rights a century ago.

Making connections

The concept of climate change is also discussed within the chapter on the carbon cycle (page 18).

Making connections

The concept of carbon budget is also discussed within the chapter on the carbon cycle (page 17).

Knowledge check 34

Explain how other negative feedback cycles could actually reduce carbon emissions from melting permafrost.

Knowledge check 35

What are the difficulties facing the peoples of the Arctic in their attempts to manage their environments at present and in the future?

Case studies

You are required to consider two case studies:

1 A case study of a glacial environment at a local scale to illustrate and analyse fundamental glacial processes and the landscapes that have been created by these processes. It is important that you engage with field data within this location. Examples in the UK include the Lake District, Snowdonia or the Highlands and isles of Scotland such as the Cairngorms or the Isle of Arran.

2 A case study of a contrasting landscape beyond the UK to illustrate and analyse how it presents challenges and opportunities for human occupation and development. You should also use this case study to evaluate human responses to the environment in terms of resilience, mitigation and adaptation. Examples could include the exploitation of oil and gas on the fringes of the Arctic Ocean, or an examination of the future of ski resorts in the Alps.

Summary

After studying this topic, you should be able to:

■ understand how glaciers and processes in cold environments operate as natural systems, and develop their own landscapes

■ describe the global distribution of past and present cold environments and outline their main physical characteristics

■ discuss the range of geomorphological processes that operate in cold environments including weathering, and erosion, transportation and deposition by the actions of ice (glacial), meltwater (fluvioglacial) and subaerial processes including periglacial

■ describe the variety of landscapes that have been created in cold environments over time, largely due to the actions of ice (glacial), meltwater (fluvioglacial) and subaerial processes, including periglacial

■ explain the varying impacts of human activity on environmentally fragile cold environments over time, including the impact of climate change

■ consider the ways in which fragile cold environments can be managed

Exam tip

The exam questions on the second case study are likely to use one or more of the following words: sustainable, resilience, mitigation and adaptation. Make sure you understand these terms.

■ Hazards

The concept of hazard in a geographical context

Nature, forms and potential impacts of hazards

A hazard is defined as a perceived natural/geophysical event that has the potential to threaten both life and property. A hazard has impacts that are social (loss of life and injury), economic (property damage, employment prospects and community loss) and environmental. A geophysical event would not be hazardous without some human occupancy of the location affected.

There are a number of different types of hazards:

- **geophysical:** all the tectonic hazards of volcanoes, earthquakes and tsunami
- **hydrological:** essentially the extremes of wet weather, i.e. droughts and floods
- **atmospheric:** tropical storms (known by a variety of names – hurricanes, typhoons, cyclones and willy-willies), tornadoes and extra-tropical storms (such as deep depressions affecting the British Isles)
- **geomorphic:** landslides and avalanches
- **biohazards:** wildfires and locust plagues
- **multiple hazardous zones:** places that experience a combination of the above

Few hazards are entirely natural. Their relationship with disaster is the result of human **vulnerability**. Some hazards, such as wildfires, can be naturally occurring, or they can be caused by the direct or indirect impact of human actions – deliberately setting vegetation alight or carelessness. In many cases, human actions intensify the impact of natural hazards, for example exacerbating earthquake risk by building inappropriate buildings.

A **disaster** is the realisation of a hazard, when it causes a 'significant impact' on a vulnerable population. It causes serious disruption of the functioning of a community or society, involving widespread/serious socio-economic and environmental losses that exceed the ability of the community (local/national) to cope. In other words, it exceeds their capacity and **resilience** level.

Risk is the probability of a hazard occurring and creating a loss of lives and/or livelihoods. It might be assumed that risk to a hazard is involuntary, but in reality many people consciously place themselves at risk. Consider, for example, all the people who live in the shadow of volcanoes. **Risk assessment** defines the likelihood of harm and damage. For this we should consider the probability of an event occurring and the severity of the hazard when it does occur. If both are high, then the disaster is likely to be greater in magnitude.

Vulnerability implies a high risk of exposure to hazards combined with an inability to cope with them. In human terms it is the degree of resistance offered by a social system to the impact of a hazardous event. In turn this depends on the resilience of the individuals and communities, and the reliability and functioning of management systems that have been put in place to deal with the event. Poverty and low economic status can amplify vulnerability.

Knowledge check 36

Distinguish between the primary and secondary effects of a hazard.

Vulnerability The risk of exposure to hazards combined with an inability to cope with them.

Resilience The degree to which a population or environment can absorb a hazardous event and yet remain within the same state of organisation, i.e. its ability to cope with stress and recover.

Knowledge check 37

Explain how the level of risk can change over time.

Hence the relationship between environmental hazards and the potential impacts on the people and areas affected is a very complex one. When examining the impact of any such event, you should always consider its magnitude, duration and extent, but also the degree to which different groups of people affected can cope and can respond. The former are inherently physical in nature, whereas the latter are more functions of the human geography of the area affected.

Hazard perception

People react to the threat of hazards in different ways because of the way in which individuals receive and process information, in turn based on their economic and cultural background. Perception is influenced by many factors including:

- socio-economic status
- level of education
- occupation and employment status
- religion
- ethnicity
- family and marital status
- past experience
- values, attitudes and expectations

Perception of a hazard will ultimately determine the course of action taken by individuals in order to modify the event or the responses they expect from governments and other organisations.

There is often a great difference in the perception of a hazard between peoples of differing levels of economic development. In wealthier areas there is a sense that the better you are prepared, the more able you will be to withstand the impact of the hazard (resilience) and perhaps even prevent the disaster from taking place. This is usually based upon government and community action, and is backed by capital that will fund technology-based solutions.

The sense of helplessness in the face of natural hazards tends to increase with the level of poverty and the deprivation of the people. Even in wealthier countries there are groups of disadvantaged people who tend to look upon natural hazards as part of their way of life as they are seen as unavoidable, just as the bulk of people in poorer countries see the impacts of these events as being part of the conditions of poverty.

The way people perceive natural hazards can be classified into the following:

- **Fatalism (acceptance):** hazards are natural events that are part of living in an area. Some communities would go as far as to say that they are 'God's will'. Action is therefore usually concerned with safety first. Losses are accepted as inevitable and people remain where they are.
- **Adaptation/adjustment**: people see that they can prepare for, and therefore survive, the event(s) by **prediction**, prevention, and/or protection depending upon the economic and technological circumstances of the area in question.
- **Fear:** the perception of the hazard is such that people feel so vulnerable to an event that they are no longer able to face living in the area, and move away to regions perceived to be safe.

Making connections

The endogenous features and character of a place, and of the people who live there, can determine both the impact, and management, of a hazard. This is discussed further within the Human geography topic of Changing places.

Adaptation/adjustment The changing of lifestyles or behaviours to cope with the threats and impacts before and after a hazardous event.

Prediction The ability to give warnings so that action can be taken to reduce the impact of hazard events. Improved monitoring and use of technology have meant that predicting hazards and issuing warnings have become more important.

Factors influencing the relationship between hazard and disaster

The factors leading to disasters can be related to both the physical profile of the hazard event and the human context in which it occurs.

Physical factors

Magnitude

Magnitude (the size of the event) is perhaps the key physical factor, but the correlation between magnitude and level of disaster is far from direct. Earthquake magnitude is now measured by the logarithmic moment magnitude scale (MMS), a modification of the earlier Richter scale. The damaging effects are measured by the Mercalli scale (useful for the impacts of shaking).

Volcano magnitude is measured by the Volcanic Explosivity Index (VEI), based on the volume and column height of ejections. This index is closely related to the type of magma that influences the type of eruption. This can be related back to the type of plate boundary the volcano is located on. Effusive eruptions of basaltic lavas with low VEI are associated with constructive boundaries or plumes, whereas explosive eruptions (with high VEI) of andesitic or rhyolitic lava are associated with destructive boundaries.

The term 'intensity' is often used when describing the magnitude of a tropical storm, and relates to the low level of atmospheric pressure within the storm.

Frequency

Frequency refers to how often an event occurs and is sometimes called the recurrence interval, such as 'a one in a 100 year event'. For most hazards there is usually an inverse relationship between frequency and magnitude. The effect of frequency on severity of impact is difficult to gauge but theoretically, areas experiencing frequent tectonic events usually have a plethora of both adaptation and **mitigation** measures, ranging from extensive monitoring, education and community awareness of what to do (for example, public education programmes, practice of evacuation procedures, the storage of emergency medical and food supplies, and planning for their delivery), to various technological strategies for shockproof/'life safe' building design (e.g. in Tokyo, Japan and San Francisco, USA) or protection (e.g. Japanese tsunami walls).

Mitigation The reduction of the amount and scale of threat and damage caused by a hazardous event.

Duration

Duration refers to the length of time that a hazard exists. Often the initial event is followed by aftershocks (e.g. Christchurch, New Zealand), or a series of subsequent eruptions (e.g. Mount Merapi, Indonesia). While individual earthquakes often last for only 30 seconds, the damage can be very extensive. Secondary hazards often prolong the duration and the damage, for example, the 'triple whammy' of the 2011 Tohoku (Japan) multi-disaster (earthquake, tsunami and nuclear accident) or the secondary hazards associated with volcanic eruptions such as lahars (e.g. Mount Pinatubo, Philippines) or jökulhlaups (glacier bursts in Iceland).

Areal extent

This is the size of the area covered by the hazard. This can have a very clear impact, as was the case in the Icelandic ash clouds after Eyjafjallajökull in 2012. These disrupted the whole of the northern hemisphere air transport system for a week, leading to widespread economic losses.

Spatial concentration

This refers to the areal distribution of types of hazards over space, such as earthquakes and volcanoes associated with types of plate boundary. In theory, hazardous regions are avoided for permanent settlement, although the opportunities provided by fertile soils encourage agricultural settlements as on the flanks of Mount Merapi and Mount Etna (Italy). Active tectonic landscapes, especially volcanic examples, also encourage tourism. In general, spatial concentration promotes sound strategies for management of the hazard, and disasters are rare.

Regularity

Regularity refers to the temporal distribution of hazards, which can add to their disaster potential. While **gap theory** can increase the possible prediction of the 'big one', earthquakes are, in reality, very unpredictable. Volcanic eruptions can be hard to predict precisely, even with close monitoring.

Human factors

Economic factors, including development

Human vulnerability is closely associated with levels of absolute poverty and the economic gap, or inequality, between rich and poor. Poverty exacerbates disasters (e.g. Haiti, Kashmir). The poorest countries lack money to invest in education, social services, basic infrastructure and technology, all of which help communities overcome disasters. Economic growth, however, increases economic assets and therefore raises potential risk levels unless managed effectively.

Social factors

World population is growing, especially in developing nations, with higher levels of urbanisation and many people living in dense concentrations of population in unsafe living settings, such as poorly sited squatter settlements. Some huge cities are very vulnerable to post-earthquake fires, as was the case in Kobe, Japan. Relief, rescue and recovery efforts are very difficult in some areas. For example, in Kashmir following the 2005 earthquake, isolation, low temperatures and the region's frontier position complicated relief and recovery. An increasingly ageing population, as in Sichuan, China, escalated vulnerability to the problems associated with evacuation and survival.

Political factors

The lack of strong central government produces a weak organisational structure. Equally, a lack of financial institutions inhibits both the disaster mitigation and emergency and post-disaster recovery. A good, strong central government leads to highly efficient rescue, as illustrated in the Sichuan earthquake, China.

Technological factors

While community preparedness and education can prove absolutely vital in mitigating disasters, technological solutions can play a major role, especially in building design, and prevention and protection.

In summary, while a hazard's physical profile can lay the foundations for the development of a disaster, it is the human profile that impacts vulnerable communities and societies, and causes disasters. The most vulnerable people,

---

Making connections

Population distribution and the factors that determine it are discussed further within the Human geography topic of Population and the environment.

Gap theory Where there has been a 'gap' in time since the last event, then it is more likely to occur in that location.

Knowledge check 38

Explain how the speed of a hazard's onset can be critical.

Making connections

Inequalities, conflicts and injustices for people and places are discussed further within the Human Geography topic Global systems.

---

such as those suffering chronic malnutrition, disease, armed conflict, chaotic and ineffective governance and lack of educational empowerment, are generally in the least resilient environments.

Characteristic human responses

Natural hazards pose a risk to human life, livelihoods and possessions. The response to hazards can occur at a variety of scales, from the individual to the local community to regional, national and international level, and, for large events, at a global scale. The choice of response depends on a complex and interlinked range of physical and human factors.

The following physical factors can affect responses:
- geographical accessibility of the location/region affected
- type of hazard, i.e. scale, impact, magnitude, frequency
- topography of the region affected
- climatic factors

The following human factors can affect responses:
- number of people involved or affected
- degree of **community preparedness/risk sharing**
- technological resources
- scientific understanding and expertise
- level of general education and training
- economic wealth of the region affected
- the quality and quantity of the infrastructure in the area, i.e. roads, railways, airports, health facilities
- the political framework (governance), i.e. competency and organisation

When taking an overview of hazardous events and their ability to develop into disasters, one approach is to develop a framework of possible strategies (Table 4).

Table 4 A hazard response framework

Strategy	Suggested policies
Modify the loss (adaptation)	■ Provide aid ■ Provide insurance
Modify vulnerability (adaptation)	■ Predict and warn (forecasting) ■ Prepare the community (risk sharing) ■ Educate to change behaviours and prevent hazards becoming disasters
Modify the event (mitigation)	■ Provide some environmental controls to reduce impact ■ Avoid hazards by land use zoning ■ Design buildings that are hazard resistant ■ Retro-fit buildings to offer some protection
Modify the cause (mitigation)	■ Have total environmental control – prevent the hazard at source (only possible for some small-scale hazards)

The choice of which of these strategies to utilise will vary during the different stages of a hazard, as shown in Park's disaster-response model (Figure 11). This is an attempt to model the impact of a disaster from before the event to after the event. It also

Exam tip

Note how this section and the following section interconnect with other parts of the specification (globalisation, governance, urbanisation, population and the character of places). Some questions will require you to make these connections and links.

Community preparedness/ risk sharing Involves prearranged measures that aim to reduce the loss of life and property damage through public education and awareness programmes, evacuation procedures, and the provision of emergency medical, food and shelter supplies.

considers the role of emergency relief agencies and rehabilitation. With each hazard event, different areas affected may have a different response curve, as the physical and human factors we saw earlier in this chapter may vary in impact.

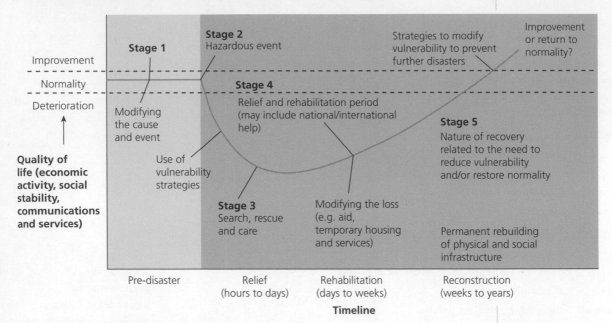

Figure 11 Park's disaster-response model

Park's model is sometimes placed in a circular format, when it is known as the hazard management cycle (Figure 12).

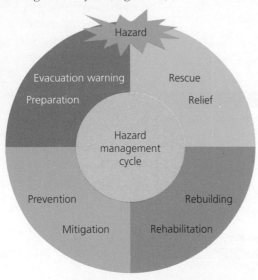

Figure 12 The hazard management cycle

Responses to hazards are controlled by the capability of the people, individuals and groups involved in the management and/or mitigation of those hazards. You should be able to consider and discuss the role of the major players in hazard management such as governments, non-governmental organisations (NGOs), insurance people, engineers and the media.

Plate tectonics

Earth structure and internal energy sources

The theory of plate tectonics suggests that the Earth's **crust** is split up into seven large rigid plates and several smaller ones, all of which are able to slowly move on the Earth's surface. They float on the semi-molten **mantle**. Geoscientists now suggest that the crust/mantle division is more complicated and have introduced new terminology to reflect this:

- the **lithosphere:** the crust and that part of the mantle just below it (80–90 km thick) which form the plates
- the **asthenosphere:** the upper part of the mantle

The centre of the Earth (the **core**) is very hot (over 5,000°C) which is caused by both primordial heat and radioactive decay. Some of this heat moves upwards into the asthenosphere.

Until the 1990s, the main plate-driving force had been thought to be mantle convection, but geoscientists have shown that two other forces should be considered:

- **Slab pull:** where plates become pulled down into the mantle (subducted) because these sections are more dense than the mantle material below, and so sink down – pulling the rest of the plate behind them (Figure 13).
- **Ridge push (gravitational sliding):** when new plates are formed at divergent plate margins in the oceans, the new plate material is hot, and less dense that the surrounding area and so it rises to form oceanic ridges. The newly formed plates slide sideways off these high areas, pushing the plate in front of them.

Most recent evidence shows that:

- Slab pull is the main plate-driving mechanism.
- Ridge push can have an effect where slab pull is not the main plate driver.
- There is little or no evidence that convection currents in the mantle move plates (apart from some very small plates in specific circumstances).

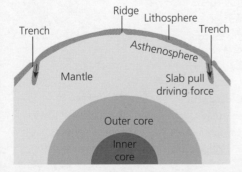

Figure 13 The structure of the Earth, and slab pull

Making connections

The importance and quality of decision-making at all levels of society are discussed further in the Human geography topics of Global systems and Global governance.

It should be pointed out that the above remains just a theory, because the logistics and cost of drilling down to the Mohorovičić discontinuity (known as 'the moho') – the boundary between the Earth's crust and the upper mantle, at its thinnest a distance of just 10 km – remain insurmountable.

Plate tectonic theory and associated landforms
Destructive plate margins

Destructive plate margins occur where two plates converge and one plate plunges under the other due to movement. At this point the plate is reincorporated into the upper mantle and crust. The area where this material is lost is called a **subduction zone**.

The three main types of destructive plate margin are as follows:

- **Oceanic/continental convergence:** here, part of the ocean floor is dragged down by a downward gravitational force (**slab pull**) beneath a continental crust, with the sinking crust forming a very deep trench, parallel to the coast. An example is the Peru/Chile trench off the west coast of South America, where the Nazca and South American plates converge. In turn, the overriding South American Plate is being lifted up, creating a chain of **young fold mountains**, the Andes. The friction between the sinking oceanic plate and the continental plate creates great heat and melting, resulting in volcanoes along the length of the mountain chain. Strong, destructive earthquakes and the rapid uplift of mountain ranges are common in this region.

- **Oceanic/oceanic convergence (island arcs):** when two oceanic plates converge, one usually sinks (subducts) under the other and in the process a trench is formed. For example, the Marianas Trench marks where the fast-moving Pacific Plate converges against the slower-moving Philippine Plate. Subduction processes in oceanic/oceanic plate convergence also result in the formation of volcanoes that, over millions of years pile up lava on the ocean floor until a submarine volcano rises above sea level to form an island volcano. Such volcanoes are typically strung out in curved chains called island arcs. The descending plate also provides a source of stress as the two plates interact, leading to frequent moderate to strong earthquakes.

- **Continental/continental convergence:** as the plates forming continental crust have a much lower density than the underlying layers, there is no subduction where they meet. Hence, as these plates move towards each other, their edges and the sediments between them are forced up to create fold mountains. There is no volcanic activity, but the movement of the plates can trigger shallow-focus earthquakes. Material can also be forced downwards to form deep mountain roots. An example of such a margin is where the Indo-Australian Plate is being forced northwards into the Eurasian Plate, creating the Himalayas.

Constructive plate margins

Constructive plate margins occur where two plates diverge away from each other allowing new magma (that formed by a reduction in pressure allowing previously solid or semi-molten mantle to melt and rise) to reach the surface. There are examples of constructive plate margins on both sea and land. For the former, mid-ocean ridges are created, whereas for the latter rift valleys occur.

An example of a mid-ocean ridge is the Mid-Atlantic Ridge (MAR). This is a submerged mountain range that runs north/south for 15,000 km through the Atlantic Ocean, from Iceland to a point 7,200 km east of southern South America. It breaks the ocean's surface in several places, forming groups of volcanic islands such as Iceland and Tristan da Cunha.

The MAR is split by a deep rift valley along its crest, 10 km wide with walls 3,000 m high on either side. It marks the boundary between the two divergent plates of the ocean's floor. This rift valley is widening at a rate of 3 cm per year. Magma from beneath the Earth's surface rises to create the high ridges on either side which, with further **gravitational sliding**, push away the lithosphere on either side (**ridge push**). This means that the rocks of the ocean floor on either side of the ridge move sideways, a process known as **sea-floor spreading**.

A rift valley system exists in East Africa where divergence has created the Great East African Rift system, which extends from the Red Sea in the north to the East African Highlands, Tanzania, in the south. These rifts are the site of some major lake basins, including lakes Malawi, Tanganyika and Edward. Individual rifts are tens of kilometres in length, up to 50 km wide and 1–5 km deep. Some geologists consider them to be incipient plate margins, and this is supported by the existence of volcanoes such as Mount Kilimanjaro.

Conservative plate margins

A **conservative plate margin** exists where two plates do not come directly into collision but slide past each other, in parallel, along a fault. The most well-known example of such a boundary is the San Andreas Fault in California, USA, which separates the northward-moving Pacific Plate (moving at 6 cm per year) and North American Plate (moving at 3 cm per year). This is a zone of intense earthquake activity, because the movement along the faults is irregular rather than a smooth process of gradual creep.

Magma plumes

The theory of plate tectonics offers an all-embracing explanation of the current distribution of the Earth's continents, volcanoes and earthquakes. There are, however, some exceptions. For example, the Hawaiian Islands are a zone of great volcanoes and yet they do not occur at the boundary of a plate. The Hawaiian Islands have formed in the middle of the Pacific Ocean more than 3,200 km from the nearest plate boundary. This island chain has resulted from the Pacific Plate moving over a deep, stationary **hotspot** in the mantle, located beneath the present-day position of the island of Hawaii (known as the 'Big Island'). Heat from this hotspot produces a constant source of magma by partially melting the overriding Pacific Plate. The magma, which is lighter than the surrounding solid rock, then rises through the mantle and crust (as a thermal **magma plume**) to erupt onto the sea floor. Over time, countless eruptions have caused a volcano to grow until it finally emerged above sea level to form an island. Continuing plate movement (from south east to north west) will eventually carry the island beyond the hotspot and volcanic activity will cease. To the far north west of the islands is a chain of underwater seamounts – the remnants of volcanic activity that took place even further back in time. This is further evidence that the plates on the Earth's crust are moving.

Knowledge check 39

Paleomagnetism provides evidence of sea-floor spreading. What is paleomagnetism, and how does it support the idea of sea-floor spreading?

Exam tip

Familiarise yourself with the names of volcanoes associated with constructive plate margins. A map showing the distribution of volcanoes will help.

Conservative (or transform) margins
Where crust is neither produced nor destroyed as the plates slide horizontally past each other.

Hotspot Area in which heat under the Earth's crust is localised. At such points, rising magma can produce volcanoes.

Volcanic hazards

Knowledge check 40

Some geologists suggest Iceland sits on a hotspot. Investigate this theory.

Vulcanicity The process through which gases and molten rock are either extruded on the Earth's surface or intruded into the Earth's crust. It is clearly linked to the existence of plate margins.

The nature of vulcanicity and its relation to plate tectonics

Volcanoes are built by the accumulation of their own eruptive products: lava, bombs (crusted-over ash deposits), and tephra (airborne ash and dust). A common form of volcano is a conical hill or mountain built around a vent that connects with reservoirs of molten rock below the surface of the Earth. There are approximately 500 active volcanoes around the world. Only a few of them are erupting at any one time. An eruption is when a volcano gives off quantities of lava, ash or volcanic gas. A few volcanoes erupt more or less continuously (e.g. Mauna Loa, Hawaii), but others lie dormant between eruptions when they give out very little gas and lava. The type of volcano and volcanic activity depends upon the nature of the lava. This in turn depends upon the location of the volcano with regard to tectonic plate margins. If the lava is a thin fluid (not viscous), then gases may escape easily. But if the lava is thick and dense (highly viscous), the gases will not move freely but will build up tremendous pressure, and ultimately escape with explosive eruptions (see Tables 5 and 6).

Table 5 Variations in the type of volcanic activity in relation to types of plate margin

	Destructive margin	Hotspot	Constructive margin
Magma source	A mix of old oceanic plate, ocean sediments and continental fragments, often weathered by water	Deep in the asthenosphere (mantle)	Deep in the asthenosphere (mantle)
Rock name	Andesite/Rhyolite	Basalt/Gabbro	Basalt/Gabbro
Magma chemistry	Medium to high acidity, greater than 63% SiO_2 (silica) content	Quite basic (alkaline), sometimes relatively rich in sodium and potassium, low silica content (around 50%)	Very basic (alkaline), low silica content, typically high iron and magnesium content
Magma's physical character	Viscous (solidifies quickly), flows over short distances, solidifies even on steep slopes, allows gases to build up pressure – can explode violently	Quite non-viscous (fairly runny), flows over low-angled slopes or can erupt as an ash	Very non-viscous (runny), flows long distances over very low-angled slopes or can create a black ash (tephra) when exploding with water vapour (steam)

Table 6 Variations in the type of volcanic activity in relation to lava type

	Basaltic lava	Andesitic lava	Rhyolitic lava
Silica content	45–50%	55–60%	65%
Eruption temperature	1,000°C+	800°C	700°C
Viscosity and gas content	Very runny, low gas	Sticky, intermediate gas	Very sticky, high gas
Volcanic products	Very hot, runny lava (shield volcanoes, low land or plateaux)	Sticky lava flows, tephra, ash, gas (composite volcanoes)	Pyroclastic flows, gas and volcanic ash (domes)
Eruption interval	Can be almost continuous, as on Hawaii	Decades or centuries	Millennia
Tectonic setting	Oceanic hotspots and constructive margins	Destructive plate margins (ocean/continental and ocean/ocean)	Continental hotspots and continental/continental margins
Processes	Dry partial melting of the upper mantle/lower lithosphere, basaltic magma is generally uncontaminated by water etc.	Wet partial melting of subducting oceanic crust contaminated by water and other material as magma rises.	In situ melting of lower continental crust, most rhyolitic (granitic) magmas cool before they reach the surface.
Hazardous?	Not really	Very	Very (but rare)

Form and impact of volcanic hazards

The impact of a volcano is only deemed hazardous when it has an effect on people. Figure 14 summarises the various forms of hazardous volcanic activity.

Exam tip

The forms of hazard associated with a volcano will vary according to the individual volcano studied. When examining your chosen case study of a recent volcanic event (see page 71), make sure you note the forms of hazard for that volcano.

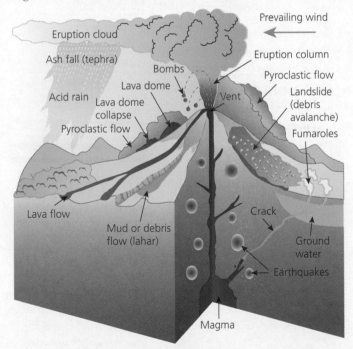

Figure 14 Forms of volcanic hazard

A volcanic event can produce a number of hazardous effects, the impact of which can range from the area immediately around the volcano to the entire planet. They include the following:

■ **Tephra:** solid material of varying grain size (from fine ash up to volcanic bombs) ejected into the atmosphere. Buildings often collapse under the sheer weight of ash falling on to their roofs. Air, thick with ash, is very difficult to breathe, and can cause serious respiratory problems. Fine tephra can also contribute to acid rain.

- **Pyroclastic flows:** very hot (800°C), gas-charged, high-velocity flows (over 200 km/h) of a mixture of gases and tephra. These flows devastate everything in their path.
- **Nuée ardente:** a glowing cloud of hot gas, steam and dust, volcanic ash and larger pyroclasts produced during a violent eruption, which can descend the slopes of a volcano at high velocity.
- **Lava flows:** at the speed at which most lava flows, they do not usually pose a threat to life. Lava flows do, however, represent a threat to farmland, property and infrastructure.
- **Volcanic gases:** these include carbon dioxide, carbon monoxide, hydrogen sulfide, sulfur dioxide and chlorine.
- **Lahars:** volcanic mud flows such as the one that devastated the Colombian town of Armero after the eruption of Nevado del Ruiz in November 1985, which buried some people alive.
- **Flooding:** caused by the melting of ice caps and glaciers, such as glacial bursts (or **jökulhlaup**).
- **Tsunamis:** giant sea waves set off by huge explosions such as the one that devastated the island of Krakatoa in 1883.
- **Climatic change:** the ejection of vast amounts of volcanic debris into the atmosphere can reduce global temperatures and is believed to have been an agent in past and present climatic change.

One way of classifying a volcano's hazardous nature is by its explosivity, using the Volcanic Explosivity Index (VEI). A volcano's impact can be judged in terms of its primary and secondary effects, and the environmental, social, economic and political consequences, both short and long term.

Management and responses

The management and responses to volcanic events fall into two categories:

- Prediction:
  - Study the eruption history of the volcano.
  - Measure gas emissions, land swelling, groundwater levels (e.g. by using GIS).
  - Measure the shock waves generated by magma travelling upwards.
- Protection:
  - Assess the hazard, i.e. try to determine the areas of greatest risk that should influence land use planning.
  - Dig trenches to divert the lava.
  - Build barriers to slow down lava flows.
  - Administer explosive activity to try to divert a lava flow.
  - Pour water on the lava front to slow it down.

Exam tip

Although you have to study one case study of a recent volcanic event in detail, be aware of the **impacts** of other volcanic events (although this doesn't require quite as much detail).

Knowledge check 41

How can geologists determine the frequency of volcanic eruptions?

Exam tip

Although you have to consider one case study of a recent volcanic event in detail, be aware of the **management of and responses to** other volcanic events (although this does not require quite as much detail).

Case study requirement

You are required to study **one recent volcanic event**, its impacts and the human responses to it. Possible volcanoes include: Mt St Helens (Washington, USA), Mt Nyiragongo (DRC), Mt Etna (Italy), Soufrière Hills (Montserrat), Mt Merapi (Indonesia) and Eyjafjallajökull (Iceland). The framework in Table 7 may assist in this process.

Note: an example of a completed table (for an earthquake) is provided on pages 74–75.

Table 7 Case study disaster framework

Case study				
Location/Date				
Geographical context (e.g. plate boundary(ies))				
Description of event	Magnitude	Duration	Linked events	Other features
Impacts	Primary		Secondary	
Environmental				
Social				
Economic				
Political				
Risk management including preparedness and prevention				
Mitigation strategies				
Adaptation strategies				

Exam tip

Although you are encouraged to keep up-to-date with events, when undertaking these case studies it is recommended that you choose disasters that have run their course and are at least 2–3 years old. In this way all the requirements can be met.

Seismic hazards
The nature of seismicity and its relation to plate tectonics

Most earthquakes occur along plate boundaries or deep under continents. Their location can also be linked to the distribution of certain geological characteristics, such as conservative plate margins and associated transform faults (low frequency but high predictability), ocean trenches and beneath mountains (greater frequency and high predictability).

As the Earth's crust is mobile, there can be a slow build-up of stress within the rocks where movement is taking place. When this stress is suddenly released, parts of the surface experience an intense shaking motion that lasts for just a few seconds. This is an earthquake.

Seismicity The geographic and historical distribution of earthquakes. Again, their distribution is closely associated with plate margins.

The depth of **focus** of an earthquake is significant:

- shallow earthquakes (0–70 km) (75% of all energy released) cause the most damage
- intermediate (70–300 km) and deep (300–700 km) earthquakes have much less effect on the surface

Earthquakes originate along faults. Parts of the crust are being forced to move in opposite directions, or in the same direction but at different speeds. These huge masses of rock get stuck, but the forces on them continue, building up stresses in the rocks. Eventually the strain overcomes the elastic strength of the rocks and they fracture, releasing large amounts of energy. At the moment of fracture the rocks regain their original shape but in a new position. The quaking and shaking takes place during the sudden movement of the rock back to its original shape, after the stress is released. This energy is transferred to the surrounding rocks, travelling through them as seismic waves. A lot of the energy is transferred vertically to the surface and then moves outwards from the **epicentre**.

Form and impact of seismic hazards

A seismic event is only deemed hazardous when it has an impact on people. It can produce a number of hazardous effects, the impact of which can range from the area close to the event to large parts of the planet. They include the following:

- **Earthquakes:** a series of vibrations and shock waves initiated by movements along the boundaries of tectonic plates.
- **Aftershocks:** smaller earthquakes that occur after a previous large earthquake in the same area.
- **Tsunamis:** large waves that flood areas along a shoreline and up estuaries, often caused by submarine earthquakes. When they are out at sea, they have a very long wavelength, often in excess of 100 km. They are very short in amplitude, at around 1 m in height. They travel very quickly, often at speeds of up to 700 km/h, for example taking less than a day to cross the Pacific. When they reach land they rapidly increase in height, up to over 25 m in some cases. They are often preceded by a localised drop in sea level (drawback) as the approaching tsunami water is drawn back and up.
- **Liquefaction:** where soils with high water content lose their mechanical strength when shaken violently during an earthquake. They behave like a fluid.
- **Landslides:** mass movement of rock down a mountainside triggered by the shaking of the ground during an earthquake.

An earthquake's magnitude is measured by a number of scales including (in order of date of development) the **Mercalli scale**, the **Richter scale** and the **moment magnitude scale (MMS)**. The Mercalli scale measures the effects of an earthquake and has a 12-point scale. The Richter scale measures the magnitude in terms of the energy released and has a 10-point logarithmic scale. Both of these have been superseded by the MMS (denoted as M_W), which also measures earthquakes in terms of the energy released. Humans rarely feel earthquakes of M_W2 or less. The scale is also logarithmic. An increase of 1 unit of magnitude increases the amount of shaking by 10, but the amount of energy released by 30.

Focus The point below the Earth's surface at which an earthquake occurs.

Epicentre The point on the Earth's surface directly above the focus of an earthquake.

Exam tip

The forms of hazard associated with an earthquake will vary according to the individual earthquake studied. When examining your chosen case study of a recent seismic event (see page 73), make sure you note the forms of hazard for that earthquake.

Knowledge check 42

Explain how an earthquake causes a tsunami.

A seismic hazard's impact can be judged in terms of its primary and secondary effects, and the environmental, social, economic and political consequences, both short and long term.

Management and responses

The management and responses to seismic events fall into three categories:

- Prediction:
  - Study groundwater levels, the release of radon gas (both using GIS) and animal behaviour.
  - Monitor fault lines and local magnetic fields.
  - Study fault lines to look for 'seismic gaps' at which the next earthquake may occur.
- Prevention:
  - Keep the plates sliding past each other rather than 'sticking' and then releasing. Suggestions include using water and/or oil.
- Protection:
  - Build hazard-resistant (aseismic) structures. For example, install a large weight that can move with the aid of a computer program to counteract stress, have large rubber shock absorbers in foundations, have cross-bracing to hold the building when it shakes.
  - Retro-fit older buildings and elevated motorways with such devices.
  - Educate people in survival strategies and encourage earthquake drills.
  - Provide signs of evacuation routes in tsunami-prone areas.
  - Advise people in assembling earthquake kits, which include bottled water, canned food, clothing/bedding, a first aid kit, torch, batteries, can opener, matches and a small fire extinguisher.
  - Install 'smart' meters that cut off gas supplies at a certain tremor threshold.
  - Maintain organisation of emergency services, ensuring the correct gear is in place (such as that required for heavy lifting).
  - Plan land use to avoid certain buildings being constructed in high-risk areas.

Exam tips

You have to consider one case study of a recent seismic event in detail, but be aware of the **management of and responses to** other seismic events (although this does not require quite as much detail).

Although you are encouraged to keep up-to-date with events, when undertaking these case studies it is recommended that you choose one which has run its course and is at least 2–3 years old. In this way all the requirements can be met.

Exam tip

Although you have to consider one case study of a recent seismic event in detail, be aware of the **impacts of** other seismic events (although this doesn't require quite as much detail).

Case study requirement

You are required to study **one recent seismic event**, its impacts and the human responses to it. Possible earthquakes include: Northridge (Los Angeles, USA), Gujarat (India), the Boxing Day tsunami at Banda Aceh (Indonesia), L'Aquila (Italy), Tohoku (Japan), Haiti, Christchurch (New Zealand), Gorkha (Nepal) and Amatrice (Italy). A completed example is shown in Table 8.

Table 8 Case study framework: earthquake (seismic hazard) example

Case study	The Sumatran-Andaman earthquake (Boxing Day tsunami)			
Location/Date	Epicentre off west coast of Sumatra/26 Dec 2004			
Geographical context	Occurred where the northward-moving (6 cm per year) Indo-Australian Plate subducts beneath the Burma Plate. There is an island arc (Andaman Islands) and a trench (Sunda Trench). Focus 30 km below the surface. There was a slip of 15 m along a 1,600 km fault.			
Description of event	**Magnitude**	**Duration**	**Linked events**	**Other features**
	$9.1–9.3\,M_W$	8–10 minutes	Indian Ocean tsunami	Daily aftershocks measuring up to $6.7\,M_W$ and lasting for 3 months
Impacts	**Primary**		**Secondary**	
Environmental	There was a 10 m movement laterally and 4–5 m vertically along the fault line. The seabed rose by several metres, which displaced 30 km³ of water. This triggered a tsunami along the whole 1,600 km length of subduction. Banda Aceh province, the land mass closest to the epicentre of the earthquake, bore the full brunt. About 15 minutes after its eruption, the tsunami hit the west coast of Aceh. In some places, waves went inland 7.5 km from the coastline. Sri Lanka was the next worst affected because there was no other landmass between it and the epicentre. The waves hit over 2,260 km of coastline in the east and north of Sri Lanka. In many areas, the walls of water were up to 10 m high when they lashed against the shoreline. In some areas waves did not exactly break, but continued inland as a fast stream of high water up to 5 km from the coast.		The smaller islands southwest of Sumatra have moved southwest by about 20 cm. Since movement was vertical as well as lateral, some coastal areas have been moved to below sea level. The Andaman and Nicobar Islands appear to have shifted south-west by around 1.25 m (and to have sunk by 1 m). The earthquake had a huge effect on the topography of the seabed. 1,500 m high thrust ridges, created by previous geological activity along the fault collapsed, generating submarine landslides several kilometres wide. One such landslide consisted of a single block of rock some 100 m high and 2 km long. The momentum of the water displaced by tectonic uplift also dragged massive slabs of rock, each weighing millions of tons, as far as 10 km across the seabed. An oceanic trench several kilometres wide was created in the seabed.	
Social	Indonesia was the worst affected area, with the death total estimated to be as high as 220,000 in Indonesia alone. Total casualties of 280,000 have been estimated. Eight people in South Africa died due to abnormally high sea levels and waves. Relief agencies reported that one-third of the dead were children. As many as four times more women than men were killed in some regions because they were waiting on the beach for the fishermen to return and looking after their children in the houses. Up to 9,000 tourists (mostly Europeans) were among the dead or missing. Sweden's death toll was 543.		1.8 million displaced people were spread over a dozen countries. There was significant loss of housing. Diseases spread, e.g. malaria, dengue fever, cholera and typhoid as well as 'Tsunami lung'. Cash-for-work programmes contributed to the reconstruction effort while providing jobs and prompting social development. Many women participating in such programmes received the same wages as men, for the first time.	

Economic		The total economic cost of damage was estimated at US$9.4 billion. In Aceh, the cost of damage (US$4.5 billion) was almost equal to its GDP in the previous year.
Political	Sri Lanka's civil war was halted temporarily as both sides helped in the rescue, recovery and the initial rebuilding phase following the tsunami.	Aceh had seen conflict between separatists and the government for 30 years. When the waves hit, fighting ceased as the parties became focused on the more immediate struggle for survival. In August 2005 a peace agreement was called for separatists to surrender their weapons and the government to withdraw its troops. This process was completed by December that year.
Risk management including preparedness and prevention	Despite a time lag of up to several hours between the earthquake and the impact of the tsunami in some parts of the Indian Ocean, nearly all of the victims were taken completely by surprise. There was no tsunami warning system in the Indian Ocean to warn the general population living around it. Tsunami detection is not easy because while a tsunami is in deep water, it has little height and a network of sensors is needed to detect it. Setting up the communications infrastructure to issue timely warnings is an even bigger problem, particularly in a relatively poor part of the world. There is a close connection between the magnitude of the damage the tsunami caused and poor coastal management. The high loss of life was partly a result of the destruction of natural defences, such as coral forests and mangrove swamps, and the building of oceanfront hotels and villas. For instance, the effects of the tsunami were less severe in areas along the east coast of Aceh, where the coastal ecosystem remained relatively untouched. In Sri Lanka, the damage was very severe in coastal areas where there had been violation of regulations prohibiting the mining of coral reefs, and destruction of coastal mangrove forests, which act as buffers against high waves.	
Mitigation strategies	A tsunami warning system became active in June 2006, following the leadership of UNESCO. It consists of 25 seismographic stations and three deep ocean sensors, which relay information to 26 national tsunami management centres.	
Adaptation strategies	The mobilisation of humanitarian aid (US$14bn) was the largest ever international response to a natural disaster on record. The number of donor countries and humanitarian organisations involved was also far greater than in any previous natural disaster. Providing jobs such as rebuilding houses and weaving rope used in fishing nets has given income to farm workers who expected to miss a year's work due to the seawater inundation that rendered their farmland temporarily useless.	

Storm hazards
The nature of tropical storms and their underlying causes

Knowledge check 43

What are the various names of tropical storms around the world?

Tropical revolving storms are systems of intense low pressure (up to about 600–700 km across) formed over tropical sea areas. They move erratically until they reach land, where their energy is rapidly dissipated. At their centre they have an area of subsiding air with calm conditions, clear skies and higher temperatures, known as the eye. They only become a tropical revolving storm when the wind speed exceeds 120 km/h. Such storms are predictable in their spatial distribution. Hurricanes in the Caribbean are also predictable in their timing and frequency – usually towards the end of summer and into autumn. They are concentrated in the tropics, specifically between 5° and 20° north and south of the equator. Once generated they tend to move westwards initially, before then switching to a more northeastwards direction as they move further away from the equator.

Tropical revolving storms begin as an area of low pressure in the tropics into which warm air is drawn in a spiralling manner. Small-scale disturbances enlarge into tropical storms with rotating wind systems, which grow into a much more intense and rapidly rotating system. A number of factors largely determine this initial formation and subsequent transition:

- There must be an oceanic location where sea temperatures are over 27°C.
- The location must be at least 5° north or south of the equator so that the effect of the **Coriolis Force (CF)** can bring about the maximum rotation of the air.
- Rapidly rising moist air (from the warm sea) cools and condenses, releasing latent heat energy, which then fuels the storm. Such storms fade and 'die' over land as the energy source is removed.
- Low-level convergence of air occurs in the lower part of the system, but this is then matched by intense upper atmosphere divergence of air, together creating an updraught of air.

There is evidence that the magnitude of tropical storms is increasing with climate change – warmer waters make them intensify more rapidly. There is also a view that their rate of movement is slower, making them more damaging. However, there is no evidence that storms are more frequent.

Form and impact of storm hazards

The hazards associated with tropical storms include:

- high winds exceeding 150 km per hour, which cause structural damage and collapse of buildings, damage to bridges and road infrastructure and loss of agricultural land
- heavy rainfall, often over 100 mm a day, which causes river flooding and sometimes landslides, particularly in areas of high relief
- storm surges result from the piling up of water by wind-driven waves and the ocean rising up under reduced atmospheric pressure – such coastal flooding can extend inland if the area near the coast is flat and unprotected

The **magnitude** of tropical storms is measured on the Saffir-Simpson scale (Table 9), which consists of five levels of central pressure, wind speed, storm surge and damage potential.

Coriolis Force (CF) The effect of the Earth's rotation on air flow. In the northern hemisphere, the CF causes a deflection in the movement of air to the right, whereas in the southern hemisphere it is to the left.

Making connections

The concept of climate change is also discussed within the chapter on the carbon cycle (page 18).

Exam tip

The forms of hazard associated with a tropical storm will vary according to the individual storm studied. When examining your two chosen case studies of recent tropical storms (see below), make sure you note the forms of hazard for those storms.

Table 9 The Saffir-Simpson hurricane wind scale

Category of storm	Wind speed (km/h)	Effect	Storm surge (m above normal water levels)
1	120–153	■ No real damage to building structures ■ Some damage to trees and vegetation ■ Some risk of coastal flooding	1.2–1.5
2	154–178	■ Some roofing material, door and window damage ■ Considerable vegetation damage	1.6–2.4
3	179–209	■ Some structural damage to small houses and utility buildings ■ Extensive coastal flooding	2.5–3.6
4	210–249	■ Extensive damage ■ Complete roof collapses possible for small houses ■ Extensive coastal erosion and flooding extending well inland	3.7–5.5
5	250+	■ Complete roof failure on many dwellings and industrial buildings ■ Major flood damage ■ Massive evacuation of residential areas may be required	Over 5.5

A storm hazard's impact can be judged in terms of its primary and secondary effects, and the environmental, social, economic and political consequences, both short and long term.

Management and responses

A number of responses and risk management strategies exist in attempting to deal with tropical storms:

- Prediction:
  - Predicting storms' origin and tracks – a form of **preparedness**. This depends on the quality of monitoring and warning systems – it is essential that warnings are as accurate as possible. Forecasting the precise power and track of a tropical storm remains problematic. The USA maintains round-the-clock surveillance of hurricanes using weather aircraft in order to increase its preparedness.
- Prevention:
  - Ongoing research into how tropical storms can be tamed through **mitigation**. Much of this effort is directed at ways of reducing the storm's energy while it is still over the ocean. One attempt has been to 'seed' the storm using silver iodide outside the eye-wall clouds. The idea is to produce rainfall, so releasing latent heat that would otherwise sustain the high wind speeds.
  - Better computer forecasting models. New technologies allow forecasters to break storms into a grid, and to use sophisticated methods to predict changes in wind speed, humidity, temperature and cloud cover. The National Oceanic and Atmospheric Administration (NOAA) in the USA has a high-resolution model to enable greater forecasting accuracy.
  - Schoolchildren in Florida practise hurricane drills, similar to those in earthquake-prone areas, as part of an awareness programme called Project Safeside.
- Protection:
  - Land use planning so that areas of highest risk have limited development, and therefore less potential economic damage.
  - Strengthening of buildings to withstand storms and floods, or erecting houses/buildings on stilts.
  - Construction of seawalls, breakwaters and flood barriers.
  - Adequate insurance before the disaster, and aid during and after the event, contribute towards modifying any loss.

The impact of tropical storms depends on a range of political and economic factors. Areas with lower levels of economic development suffer from a lack of insurance, poor land use planning, inadequate warning systems and defences, and poor infrastructure and emergency services. This usually results in a higher death toll. Even within a wealthy country such as the USA, Hurricane Katrina exposed the problems of a largely uninsured and relatively poor population who struggled to cope during, and especially after, the event.

Case study requirement

You are required to study **two recent tropical storms in contrasting areas**, their impacts and the human responses to them. Possible storms include: Cyclone Nargis (Myanmar), Hurricane Katrina (USA), Typhoon Haiyan (Philippines), Hurricane Sandy and Hurricane Matthew (both the Caribbean and USA). Use a table similar to that on page 71 for each tropical storm event.

Knowledge check 44

Why is it essential that storm warnings are accurate?

Making connections

Coastal management is discussed further within the Physical geography option Coastal landscapes.

Exam tip

Although you are encouraged to keep up-to-date with events, when undertaking these case studies it is recommended that you choose disasters that have run their course and are at least 2–3 years old. In this way all the requirements can be met.

Fires in nature
The nature of wildfires and their underlying causes

Wildfires, commonly known as bushfires (Australia) or brush fires (USA/Canada), are a normal occurrence in many ecosystems. It is important to distinguish between managed fires and wildfires. **Managed fires** are frequently used in conservation areas such as wildernesses (e.g. New South Wales, Australia) and remote national parks (e.g. Yellowstone, USA) as a necessary and beneficial tool of ecosystem management.

In contrast, **wildfires** are human-induced fires that have gone out of control, and can no longer be classed as managed, or have been deliberately started with malicious intent or through carelessness. Many wildfires occur naturally too, the result of lightning strikes, though it is estimated that lightning causes only 10% of wildfires.

The nature, intensity and rate of spread of a wildfire depends on the types of plants involved, the topography, the strength and direction of the winds and the relative humidity of the air in the region. Some fires travel close to the ground, others spread via the canopies of tall trees. Wildfires are particularly associated with areas experiencing semi-arid climates where there is enough rainfall for vegetation to grow and provide a 'fuel', yet with a dry season to promote conditions for ignition.

Wildfires are therefore concentrated in parts of Australia (New South Wales and Victoria), Canada (British Columbia), the USA (California and Oregon), South Africa and southern Europe (Mediterranean areas). Some geographers have suggested that such fires have become more prevalent in these areas as there is more movement of people into the countryside, as well as an increased incidence of drought due to climate change.

Traditionally, wildfires have not been associated with areas of tropical rainforest because of the high humidity and all-year-round rainfall. However, the burning of the rainforest in Indonesia, for clearance and forest mismanagement by logging companies, combined with the drying effect of El Niño events, have revised this view. Indeed, the combination of drying winds and the burning of relatively moist timber has led to particularly smoky wildfire-related events in parts of southeast Asia in recent years. Toxic hazes from such fires, combined with other forms of air pollution, have been an increasingly common phenomenon in cities such as Kuala Lumpur and Singapore.

Impacts of wildfires

General impacts of wildfires include:

- loss of timber, livestock and crops
- loss of valuable plant species and the creation of areas dominated by fire-resistant scrub
- damage to soil structure and nutrient content
- loss of wildlife, of particular concern when rare or endemic species are involved
- in the case of extensive fires, the loss of vegetation can lead to an increased risk of flooding

Making connections

Wildfires play a role in adding carbon dioxide to the atmosphere. This is linked to the topic of the carbon cycle (page 14).

Knowledge check 45

Why are managed fires used in some areas?

Making connections

Vegetation type is discussed further within the Physical geography option Hot desert landscapes (page 23).

Knowledge check 46

Some plants are pyrophytic. What does this mean, and can you give some examples?

Exam tip

The precise impact of a wildfire will vary according to the individual fire studied. When examining your chosen case study of recent wildfire event, make sure you note the **impact** of that event.

- temporary evacuation usually requires emergency aid for the areas affected
- property loss is increasing as a result of settlement expansion into at-risk areas
- release of toxic gases and particulate pollution
- risk to firefighters (although loss of life in general is quite low)
- heavy impact on emergency services (huge costs and large numbers of people involved controlling the outbreak)

The impact of a wildfire can be judged in terms of the fire's primary and secondary effects, and the environmental, social, economic and political consequences, both short and long term.

Management and responses

A number of responses and risk management strategies exist in attempting to deal with wildfires:

- Prevention:
  - In many countries such as Australia, Greece and Spain, the main approach to fire management has been to extinguish all fires as they occur, especially in populated areas or near to high-value timber reserves (mitigation). Management to modify wildfires also concentrates on reducing or eliminating the fuel supplies from the potential path of the fire – by controlled burning. This practice is not only controversial, but also risky.
  - Technology (including GIS) to warn areas at risk. Aircraft and satellites are used to carry out infra-red sensing to check surface ground temperatures and signs of eco-stress from desiccation.
- Protection:
  - Community preparedness leading to early warning through the use of fire towers. Citizens can be trained to act as auxiliary firefighters, to organise evacuation and coordinate emergency firefighting.
  - Education concerning home safety in high-risk areas. Supplies of fuel should be reduced, wood stores stacked correctly and adequate water hose and ladders should be available. Householders are also advised to remove dead leaves from gutters. School education concentrates on ensuring young people understand the dangers of arson and casual cigarette use, and the need to adhere to barbeque laws.
  - Land use planning (adaptation). Risk management identifies areas of high vulnerability, and planning legislation ensures houses are built in low-density clusters with at least 30 m set back from any forested area. New developments can be designed with fire breaks and wide roads for access of firefighting equipment.
  - Fire-resistant housing design is increasingly important in at-risk areas.
  - Insurance is another option, although expensive and difficult to obtain in fire-prone areas.

Exam tip

You have to consider one case study of a recent wildfire event in detail, but be aware of the management of and responses to other wildfire events (although this does not require as much detail).

Exam tip

Although you are encouraged to keep up-to-date with events, when undertaking these case studies it is recommended that you choose disasters that have run their course and are at least 2–3 years old. In this way all the requirements can be met.

Case study requirement

You are required to study **one recent wildfire event**, its impacts and the human responses to it. Possible wildfires include: southern and eastern Australia (2019), Indonesia (2014–15), California and Oregon (2018–20). Use a table similar to that on page 71.

Further case study requirements

You are required to study **one multi-hazardous environment beyond the UK** to illustrate and analyse the nature of the hazards and the social, economic and environmental risks presented, and how human qualities and responses such as resilience, adaptation, mitigation and management contribute to its continuing human occupation.

Many of the rapidly growing mega-cities of the developed and developing world are located in hazard-prone areas and are possible areas for study. In some cases, rapid urbanisation has destroyed ecosystems by deforestation, which increases the risk of flash floods. Several such cities are in close proximity to plate boundaries and the tracks of tropical storms, and within dry but forested, environments. Some are at risk of a combination of these hazards. With high population densities, hazard management in large urban areas is both expensive and complex, making disasters inevitable both socially (high concentrations of vulnerable people) and economically (huge investment in infrastructure). Such urban areas include Los Angeles, Tokyo, Bangkok, Manila and Calcutta.

You are also required to study **a specified place (at a local scale) in a hazardous setting** to illustrate the physical nature of the hazard, and analyse how the economic, social and political character of its community reflects the presence and impacts of the hazard, and the community's response to the risk. You could choose to consider another case study of an area affected by a volcano or an earthquake or a tropical storm. However, you might want to consider making use of one of your earlier case studies – just make sure it is at a relatively smaller scale. An individual affected city would satisfy this requirement – Kobe (Japan) or Christchurch (New Zealand) or New Orleans (USA) would be ideal.

Summary

After studying this topic, you should be able to:
- understand the concept of a hazard in a geographical context and appreciate the variety of forms, impacts and responses that exist in the study of hazards. On the other hand, a model of hazard management for all has been developed
- know and understand the theory of plate tectonics and understand how it and associated processes explain the formation and characteristics of a variety of landforms
- explain how volcanic hazards are created, and be able to discuss their impacts and the variety of responses to them, with particular emphasis on one case study
- explain how seismic hazards are created, and be able to discuss their impacts and the variety of responses to them, with particular emphasis on one case study
- explain how tropical storm hazards are created, and be able to discuss their impacts and the variety of responses to them, with particular emphasis on two case studies
- explain how wildfire hazards are created, and be able to discuss their impacts and the variety of responses to them, with particular emphasis on one case study
- understand how some areas of the world suffer from a combination of hazards, and recognise that a range of management strategies are therefore required in those areas

Questions & Answers

About this section

In this section of the book, two sets of A-level questions on each of the content areas are given. For each of these, the style of questions used in the examination papers has been replicated, with a mixture of short answer questions, data response questions, data stimulus questions and extended prose questions. Other than the short, knowledge-based questions, all will be assessed using a 'levels of response' mark scheme to a maximum of four levels.

The sections are structured as follows:

- sample questions in the style of the examination
- mark schemes in the style of the examination
- example student answers at an upper level of performance
- examiner's commentary on each of the above

For A-level geography, all assessments will test one or more of the following Assessment Objectives (AOs):

- **AO1**: demonstrate knowledge and understanding of places, environments, concepts, processes, interactions and change, at a variety of scales.
- **AO2**: apply knowledge and understanding in different contexts to interpret, analyse and evaluate geographical information and issues.
- **AO3**: use a variety of relevant quantitative, qualitative and fieldwork skills to: investigate geographical questions and issues; interpret, analyse and evaluate data and evidence; construct arguments and draw conclusions.

All questions that carry a large number of marks require students to consider connections between the subject matter and other aspects of geography, or to develop deeper understanding, in order to access the highest marks. The former used to be referred to as **synopticity**, but the new term is **connections** – so try to think of **links** between the subject matter you are writing about and other areas of the specification. Some questions will target specific links.

For the Physical Geography examination paper 1, two **core topics** on each paper (The Water and Carbon cycles, and one landscape system – either Hot desert or Coastal or Glacial) are covered, each worth 36 marks, and the breakdown of the questions per topic is:

- one 4-mark question (AO1) – point marked
- one 6-mark question with data – marked to two levels (AO3) – **response**
- one 6-mark question with data – marked to two levels (AO1/AO2) – **stimulus**
- one 20-mark question requiring an extended prose response marked to four levels (AO1/AO2)

There are also the **options**: the section testing **Hazards** is worth 48 marks and the breakdown of questions per topic is:

■ one 4-mark question (AO1) – point marked
■ one 6-mark question with data – marked to two levels (AO3) – **response**
■ one 9-mark question (sometimes with data: **stimulus**) requiring an extended prose response marked to three levels (AO1/AO2)
■ one 9-mark question requiring an extended prose response marked to three levels (AO1/AO2)
■ one 20-mark question requiring an extended prose response marked to four levels (AO1/AO2)

Note that the latter two questions may have an explicit connection to another part of the specification.

You should allocate 1.25 minutes per mark to answer the questions.

Note: this book does *not* cover the option Ecosystems under stress.

For each question in this book, one answer has been provided towards the upper end of the mark range. Study the descriptions of the 'levels' given in the mark schemes carefully and understand the requirements (or 'triggers') necessary to move an answer from one level to the one above it. You should also read the commentary with the mark schemes to understand why credit has or has not been awarded. In all cases, actual marks are indicated.

Question types

Short answer questions

These questions assess AO1 only, and carry 4 marks each. You are expected to provide four clear statements (though not necessarily four sentences) which address the question and demonstrate your knowledge of the topic. Your statements may provide examples, or elaboration, but these are not required.

You have about 5 minutes to answer these questions, so your points should be punchy and to the point.

Data-based questions

The examination papers have two types of questions based on data.

Data response questions

These all carry 6 marks and assess AO3 only. The command word 'analyse' is frequently used.

In general, simple or obvious statements will access Level 1; more sophisticated statements will access Level 2. Note that knowledge is not required here, so do not try to explain the data – although you may be correct in such statements, you will *not* gain any credit for them.

Here are some general tips about addressing such questions:

- consider patterns/ranges/trends
- identify anomalies/countertrends
- manipulate the data (e.g. calculate percentages, or fractions; use qualitative descriptive words) – don't simply 'lift' or copy them
- make connections/draw relationships between the different sets of data provided
- be prepared to question and/or criticise such relationships, or indeed the data provided

You have about 8–9 minutes to answer these questions.

Data stimulus questions

These questions assess AO1 and AO2 in the proportion of 2/4 for 6-mark questions (core topics) and 4/5 for 9-mark questions (optional topics). You should demonstrate that you know the factors that underpin the context of the data provided, but also apply this knowledge to the question given. It is important that you refer to the data provided, but then use it as a stimulus to answer the question. In other words, use the data provided as a 'springboard'. Note the questions often use the phrase '… and your own knowledge'. There is also often an element of assessment or evaluation in the question (such as the use of the command word 'assess') – this is the AO2 part of the question, and it must be addressed.

You have about 8–9 minutes (6 marks), or 10–12 minutes (9 marks) to answer these questions.

The 9-mark extended prose questions

These questions assess AO1 and AO2 in the proportion of 4/5 and are *only* found within the **Optional** elements. You should demonstrate that you know the factors that underpin the context of the question, but also then apply this knowledge to the question given. There is usually an element of assessment or evaluation in the question – this is the AO2 part of the question, and it must be addressed.

At least one of these 9-mark questions across the examination as a whole will connect the study of geography across this specification (sometimes referred to as the synoptic question). There is no pattern as to where such questions fall – they may be on either Paper 1 or Paper 2, or both.

You have about 12–13 minutes to answer these questions. You should aim to write about 250–350 words.

The 20-mark essay questions

You are required to write *three* evaluative essays on each examination paper. Each essay should be completed in around 25 to 30 minutes. In general, this represents around 2–3 pages of average-sized handwriting, i.e. somewhere between 400 and 600 words. These evaluative essays should incorporate an introduction and a formal conclusion, with several paragraphs (3–5) of argument in between.

These essays are assessed using a generic mark scheme such as the one below. Study this carefully to see what is needed to move from one level to the next.

Level/mark range	Criteria/descriptor
Level 4 (16–20 marks)	■ Detailed evaluative conclusion that is rational and firmly based on knowledge and understanding, which is applied to the context of the question (AO2). ■ Detailed, coherent and relevant analysis and evaluation in the application of knowledge and understanding throughout (AO2). ■ Full evidence of links between knowledge and understanding and the application of knowledge and understanding in different contexts (AO2). ■ Detailed, highly relevant and appropriate knowledge and understanding of place(s) and environments used throughout (AO1). ■ Full and accurate knowledge and understanding of key concepts and processes throughout (AO1). ■ Detailed awareness of scale and temporal change, which is well integrated where appropriate (AO1).
Level 3 (11–15 marks)	■ Clear, evaluative conclusion that is based on knowledge and understanding, which is applied to the context of the question (AO2). ■ Generally clear, coherent and relevant analysis and evaluation in the application of knowledge and understanding (AO2). ■ Generally clear evidence of links between knowledge and understanding and the application of knowledge and understanding in different contexts (AO2). ■ Generally clear and relevant knowledge and understanding of place(s) and environments (AO1). ■ Generally clear and accurate knowledge and understanding of key concepts and processes (AO1). ■ Generally clear awareness of scale and temporal change, which is integrated where appropriate (AO1).
Level 2 (6–10 marks)	■ Some sense of evaluative conclusion partially based upon knowledge and understanding, which is applied to the context of the question (AO2). ■ Some partially relevant analysis and evaluation in the application of knowledge and understanding (AO2). ■ Some evidence of links between knowledge and understanding and the application of knowledge and understanding in different contexts (AO2). ■ Some relevant knowledge and understanding of place(s) and environments, which is partially relevant (AO1). ■ Some knowledge and understanding of key concepts, processes and interactions, and change (AO1). ■ Some awareness of scale and temporal change, which is sometimes integrated where appropriate; there may be a few inaccuracies (AO1).
Level 1 (1–5 marks)	■ Very limited and/or unsupported evaluative conclusion that is loosely based upon knowledge and understanding, which is applied to the context of the question (AO2). ■ Very limited analysis and evaluation in the application of knowledge and understanding; lacks clarity and coherence (AO2). ■ Very limited and rarely logical evidence of links between knowledge and understanding and the application of knowledge and understanding in different contexts (AO2). ■ Very limited relevant knowledge and understanding of place(s) and environments (AO1). ■ Isolated knowledge and understanding of key concepts and processes (AO1). ■ Very limited awareness of scale and temporal change, which is rarely integrated where appropriate; there may be a number of inaccuracies (AO1).
Level 0 (0 marks)	■ Nothing worthy of credit.

Command words

Command words are the words and phrases used in exams and other assessment tasks that tell students how they should answer the question. The following high-level command words could be used:

Analyse (when used for extended prose questions) Break down concepts, information and/or issues to convey an understanding of them by finding connections and causes and/or effects.

Analyse (when used for 6-mark AO3 questions) In this context, this command requires students to interface with the data and deconstruct the information. (See Data response questions – page 82.)

Assess Consider several options or arguments and weigh them up so as to come to a conclusion about their effectiveness or validity.

Compare Describe the similarities and differences of at least two phenomena.

Evaluate Consider several options, ideas or arguments and form a view based on evidence about their importance/validity/merit/utility.

Examine Consider carefully and provide a detailed account of the indicated topic.

Explain/Why/Suggest reasons for Set out the causes of a phenomenon and/or the factors which influence its form/nature. This usually requires an understanding of processes.

Interpret Ascribe meaning to geographical information and issues.

To what extent Form and express a view as to the merit or validity of a view or statement after examining the evidence available and/or different sides of an argument.

■ Questions: Set 1

Water and carbon cycles

Question 1

Explain two ways in which overland flow is created. (4 marks)

> 1 mark per valid point.

Student answer

Overland flow can be created by either infiltration excess overland flow, or saturation excess overland flow ✓. In the former, overland flow occurs when the rainfall intensity exceeds the infiltration capacity of the soil. A layer of water forms on the soil surface and accumulates in small depressions, or puddles, which, when full, begin to overflow ✓.

Saturation excess overland flow occurs where the ground becomes more saturated such as the base of slopes ✓. Soils tend to be wetter at the base of a slope, nearer the stream channel. If the soil becomes totally saturated, then any extra rainwater must flow over the soil surface into the nearby stream ✓.

4/4 marks awarded Four correct statements are given.

Question 2

Figure 1 shows actual and predicted carbon dioxide emissions per capita for a selection of countries and the world between 1965 and 2035. Analyse Figure 1. (6 marks)

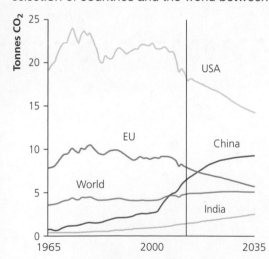

Figure 1 Carbon dioxide emissions from energy use per capita – selected countries and world 1965–2035

Level 2 (4–6 marks)

AO3 – Clear analysis of the quantitative evidence provided, which makes appropriate use of evidence in support. Clear connection(s) between different aspects of the evidence.

Level 1 (1–3 marks)

AO3 – Basic analysis of the quantitative evidence provided, which makes limited use of evidence in support. Basic connection(s) between different aspects of the evidence.

Student answer

The striking feature of the information shown in Figure 1 is that the declines that have taken place, and will continue to take place, in the amount of carbon dioxide emissions per capita from the USA and the EU are almost totally matched by the increases in India and China. The outcome of this is that the world's total emissions per capita of carbon dioxide are only predicted to rise by a small amount from 4 to 5 tonnes.

China's per capita emissions of carbon dioxide are predicted to rise rapidly to almost 10 tonnes per capita, and this is due to the ongoing industrialisation of the country making use of cheap coal and coal-fired power stations. Similarly, in India, there are signs of increase, though not as dramatic as in China.

The USA's per capita emissions are far greater than anyone else's – being at about 20 tonnes and over per capita for the last 40–50 years. This is due to their excessive use of gas-guzzling vehicles, and heavy industries pumping out lots of carbon into the atmosphere. However, the emissions are predicted to fall to 15 tonnes per capita by 2035.

Finally, the EU's emissions of carbon dioxide per capita are also predicted to fall to about 6 tonnes. This is most likely to continue due to these countries seeking to cut their emissions as they are introducing more renewable sources of energy to cut down on emissions.

6/6 marks awarded The answer begins with a sophisticated overview of the data, as well as recognising that some of the data are in the past, and some in the future. There is also some qualitative and quantitative use of the data, with clear connections made between five lines on the graph.

This paragraph contains both good description and valid analysis, again with a clear connection/comparison between India and China.

Similarly there is accurate description and analysis regarding the USA.

The final paragraph is perhaps the weakest, but follows the same themes – the student probably felt he/she needed to refer to all elements of the figure, but this was not necessary as maximum credit was already achieved.

Question 3

Figure 2 shows the global carbon cycle. Using Figure 2, and your own knowledge, assess the role of oceans in regulating the carbon cycle.

(6 marks)

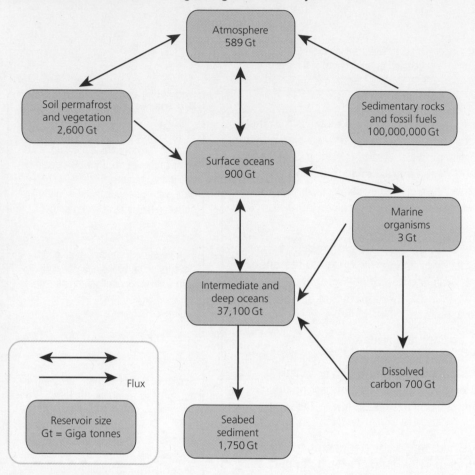

Figure 2 The global carbon cycle

Level 2 (4–6 marks)

AO1 – Demonstrates clear knowledge and understanding of concepts, processes, interactions and change.

AO2 – Applies knowledge and understanding to the novel situation offering clear evaluation and analysis drawn appropriately from the context provided. Connections and relationships between different aspects of study are evident with clear relevance.

Level 1 (1–3 marks)

AO1 – Demonstrates basic knowledge and understanding of concepts, processes, interactions and change.

AO2 – Applies limited knowledge and understanding to the novel situation offering only basic evaluation and analysis drawn from the context provided. Connections and relationships between different aspects of study are basic with limited relevance.

Student answer

The oceans act as a store of carbon. The total ocean store is over 38,000 Gt making it a vital component of the carbon cycle and so maintain planetary health. Another key role of the oceans are their ability to absorb carbon dioxide into the surface ocean store (900 GtC). Crucially, this means that they are considered a carbon sink and so highlights the importance of the oceans in regulating the carbon cycle.

There are two key processes that occur in oceans that allow them to be a carbon sink with the atmosphere. Dissolved carbon dioxide is taken from the surface ocean to the intermediate and deep oceans through down-welling currents. Carbon dioxide is also sequestered in the ocean through photosynthesis by phytoplankton and other marine biota which converts the carbon dioxide into organic matter.

Another key role of the oceans is to act as a biological pump transporting carbon from the surface oceans to the intermediate and deep oceans (37,100 Gt). This occurs when as biological organisms die, their dead cells, shells and other parts sink into the mid and deep water. Decay of these organisms then releases carbon dioxide into the intermediate and deep water (dissolved carbon 700 Gt) stores.

Some material sinks right to the bottom of the ocean and forms the seabed sediment store (1,750 GtC) where over time, through chemical and physical processes, the carbon is transformed to rocks such as limestone. The carbon sequestered in this store can remain for geological epochs and so regulates the carbon cycle.

Overall, the oceans play a vital role in regulating the carbon cycle by being the largest non-geologic store of carbon as well as being a carbon sink. However, the ability of the oceans to offset the impact of burning fossil fuels is small and if fossil fuel use continues to accelerate the oceans will have an increasingly smaller role to play in maintaining the carbon cycle.

6/6 marks awarded The student begins by making clear reference to Figure 2, and explaining in general the significance of oceans as a carbon store and sink. More detailed analysis then takes place, demonstrating both clear knowledge and application of that knowledge in the context of the novel context. Several statements in the second, third and fourth paragraphs illustrate these aspects. The answer ends with a brief conclusion, summarising the previous points, but also raising an important qualification in the overall assessment.

Question 4

Evaluate the relative importance of natural variation and human impact on changes in the water cycle over time.

(20 marks)

See the generic mark scheme on page 84.

Student answer

One natural variation I have studied is a storm event which may create flooding. Flooding is a natural occurrence which is solely dependent on the weather. The most obvious cause of flooding is a high amount of precipitation that is greater than the rate of interception and can lead to flash floods. A period of precipitation over a long time can also lead to a soil moisture surplus in the water budget or saturation of the ground where the ground stops absorbing water and therefore leads to overland flow and then flooding. Other things such as impermeable rock in the ground or melting ice and snow can influence floods. Also weather previous to an event of heavy precipitation can influence flooding. Very cold weather can cause the ground to become frozen and act like impermeable rock which will make infiltration impossible leading to a large amount of surface runoff. The natural geography of areas will impact flooding with some areas of low-lying land such as Bangladesh being prone to natural flooding where large rivers and tributaries meet and overflow on to flood plains. Here the monsoon season is also a major flood cause, as in the monsoon season much rain falls on these large flat drainage basins. In 2007 Bangladesh suffered some of the worst river floods in years – 169.5 mm of rain fell over 36 hours and this meant the ground was saturated and so excess water flowed straight to the River Ganges as surface runoff.

17/20 marks awarded All elements of the mark scheme for Level 4 have been addressed.

Human impacts are often seen as more important in the causes of flooding. However, flooding still took place before any human influence. A major cause of flooding is deforestation where humans cut down trees and vegetation either for timber resources or to build upon. This reduces the amount of interception which would usually lead to some evaporation. Also trees and vegetation would absorb a large amount of soil moisture and would therefore reduce the likelihood of flooding by taking in the water and transpiring it back into the atmosphere. Humans build impermeable surfaces with roads and areas of concrete. These are essentially very similar to frozen ground in areas of permafrost such as the Alaskan North Slope, or natural impermeable rock because they prevent infiltration.

A link is made to climate, and to cold environments – a connection to a different context.

There is a clear account of a range of both natural variation and human impacts on the water cycle.

In places such as Carlisle which lies on a floodplain the building of roads and concrete structures has resulted in severe flooding. The build-up of impermeable surfaces such as concrete pavements means that when precipitation occurs infiltration is not possible and rainwater flows straight to the river channel as surface runoff, decreasing lag time and river discharge – effectively causing a river to flood.

There is also evidence of sequencing. Some references to case studies are evident too (Bangladesh and Carlisle).

Efficient drainage systems also don't help as a human cause. Water is carried straight to the river via drains instead of gradually getting there via baseflow. An example of this was again found in Carlisle in 2005 when a large proportion of the flooding was caused by Carlisle's drainage system.

Overall the importance of both of these elements varies according to location. It is true though that human causes can be minimalised or prevented, whereas physical variations are much more dependent on nature and difficult to control, if at all. So, in relative terms, natural variation seems to be the main factor.

Explicit assessment of relative importance is given in the conclusion. However, the answer is dominated by flooding, with a small section on land use change (deforestation). A little more on other aspects of the water cycle, such as farming practices and water abstraction would have achieved maximum marks.

Hot desert systems and landscapes

Question 1

Explain two ways in which wind erosion takes places in deserts. (4 marks)

1 mark per valid point.

Student answer

The wind is a major cause of erosion in desert landscapes as it picks up sand particles and uses them to sandblast away at rocks in the area ✓. Softer rocks are more prone to erosion in this way ✓, though most of the erosion tends to take place within a metre of the ground surface ✓.

The sheer force of the wind can also pick up loose, fine material and transport it well away from its source ✓. This leaves behind a bare rock surface, or pavement.

4/4 marks awarded Four correct statements are provided. Note that the second and third statements are development points of the initial statement. This is perfectly acceptable as long as two 'ways' are given in the answer as a whole.

Question 2

Study Table 1. Calculate each of the range of average day temperatures and the mean value of average night temperatures, and analyse how the data in Table 1 support the view that Las Vegas is located in a hot desert area. (6 marks)

Table 1 Climate statistics for Las Vegas, USA

Month	Jan	Feb	Mar	Apr	May	Jun	Jul	Aug	Sept	Oct	Nov	Dec
Average day temperature (°C)	14	17	21	26	31	37	40	39	34	27	19	14
Average night temperature (°C)	3	5	8	12	17	22	26	25	20	14	7	3
Precipitation (mm)	15	17.5	15	3.8	6.1	2	11.2	11.4	7.9	6.1	7.9	10.2
Average number of days with rain	3.4	3.5	3.6	1.8	1.6	0.7	2.6	3.0	1.9	1.8	1.8	2.9
Sunshine hours	244.9	248.6	313.1	345	387.5	402	390.6	368.9	336	303.8	246.0	235.6

1 mark for each of the calculations; and then 1 mark for each valid point of analysis.

Student answer

The range of average day temperatures is 26°C ✓.

The mean of average night temperatures is 13.5°C ✓.

The city of Las Vegas is situated in a desert climate as precipitation is only 114 mm a year ✓ and its maximum temperatures can be as high as 40°C ✓. Average temperatures are 26°C, though the temperature range is also quite high, both day and night, both over 20°C ✓. Temperatures at night in winter can drop as low as 3°C. The number of days with rain is low, at less than 30 days ✓. Sunshine amounts are high all year round, especially in the summer, and the total amount of sunshine hours for the year is very large ✓.

These facts all support the view that Las Vegas is in a hot desert environment.

> **6/6 marks awarded** The student gains 2 marks for the correct calculations, and then 4 more marks (max) are awarded for analysis.

Question 3

Figure 3 shows vegetation in a hot desert environment. Using Figure 3, and your own knowledge, assess how such vegetation is adapted to low levels of precipitation. (6 marks)

Figure 3 A landscape in Oman

Level 2 (4–6 marks)

AO1 – Demonstrates clear knowledge and understanding of features, concepts, processes, interactions and change.

AO2 – Applies knowledge and understanding to the novel situation offering clear evaluation and analysis drawn appropriately from the context provided. Connections and relationships between different aspects of study are evident with clear relevance.

Level 1 (1–3 marks)

AO1 – Demonstrates basic knowledge and understanding of features, concepts, processes, interactions and change.

AO2 – Applies limited knowledge and understanding to the novel situation offering only basic evaluation and analysis drawn from the context provided. Connections and relationships between different aspects of study are basic with limited relevance.

Student answer

The vegetation shown in Figure 3 is a bush with small leaves (and possibly spikes to keep away animals) to reduce transpiration. It has a lot of thin stems, or branches. There seems to be a larger tree-like version of the plant in the right background. Such vegetation is extremely sparse with a large area of land, mostly sand dunes, containing no plants at all.

Due to the lack of precipitation here, xerophytic plants (plants that can store and retain water) like this one have an extended root system which allows them to tap into the water table while also covering a wide area of land. They have waxy cuticles which prevents water loss through transpiration. They either have small or no leaves to also reduce water loss. The spikes found on xerophytic plants mean they can store water as they may be exposed to long periods without rainfall.

Ephemeral plants in such an environment are short lived. Their seeds are able to lie dormant for months and years. However, when it does rain they are able to germinate and produce seeds within days or a few weeks.

Lastly halophytes have adapted to withstand saline conditions, due to high rates of evaporation, often found in deserts. These plants have adapted to have filter systems in the root so only water can be taken up. Some plants also store water and use it to weaken the salinity in the plant when they are mixed together.

So, it can be seen that a wide range of vegetation types have adapted to the low levels of precipitation found in areas such as that shown in Figure 3.

6/6 marks awarded The answer begins by making clear reference to the data provided, describing what can be seen, and making a brief link to reducing the level of transpiration. This first paragraph is a good start for this type of question.

The subsequent paragraphs address the question in more detail, with clear explanations for each of xerophytic plants, ephemeral plants and halophytes.

The brief conclusion addresses the assessment element of the question, with an explicit link back to the data.

Question 4

Evaluate the impact of desertification on ecosystems, landscapes and people in one area you have studied.

(20 marks)

See the generic mark scheme on page 84.

Student answer

The Sahel is an area where the ecosystem is suffering from a lack of regular rainfall. The plants that grow are increasingly drought tolerant with desert grasses and ephemeral flowers taking over larger areas. To make things worse, trees and large bushes are also being reduced in number due to the activities of people chopping them down for firewood creating a barren landscape – a key cause of desertification.

17/20 marks awarded The introduction identifies the Sahel region of Africa as the chosen area of study. The scene is set for the essay and includes a reason for desertification occurring.

Living in the Sahel of Africa is very challenging for people. Food production, from crops and animals, is both difficult and quite unsustainable. The desertification of the Sahel has proven how hard such an area is to sustain with soil erosion making it hard to grow crops, and a lack of water and vegetation making it hard to graze livestock. These often increase desertification anyway through overgrazing. Rainfall in the Sahel has decreased since the 1960s making drought a huge problem but with population growing at 2.5% per year, desertification has become an even bigger problem. Rainfall is infrequent in the Sahel and this has a number of impacts on the people who live there. Drought is not uncommon – people often have to walk miles each day to find safe drinking water.

This second paragraph features the impact of desertification on people. Here there is some sense of analysis of the problems, linking physical and human aspects.

Desertification means that soils are not suitable for growing crops and so food is not always available. A lack of food, leading to undernutrition, means that diseases are caught more easily due to a weak immune system. Many farmers are subsistence farmers and so if a crop fails, they have nothing to eat. A lack of crops means that their animals have nothing to graze on. In many sub-Saharan African countries, herd size is a sign of wealth and so a large herd is advantageous.

The discussion of problems faced by people is extended, with other cultural and physical aspects being discussed, one of which is wealth or development – a synoptic link to global systems.

Another problem faced by people in the Sahel is a fuelwood crisis. This results in a large number of trees being chopped down to provide fuelwood. However, this is unsustainable and increases further the risk of soil erosion, worsening problems and making the landscape even more barren.

This paragraph continues the landscape theme, with reference to deforestation and soil erosion.

However, aid is available for this problem and self help initiatives for the population can improve life and sustainability. Projects such as PAF, an agro-forestry project in Burkina Faso, are aiming to help the lives of locals and the sustainability of the land. The intention is to get the community involved and doing rather than the project doing it for them. Training was given as well as food and tools and resources to help them get a head start with the project. The first idea of tree planting (to reduce soil erosion) wasn't accepted by locals and so it failed. PAF then changed it slightly so that the community could grow crops. Traditional 'stone bunds' were put in lines along contours and water from the few rainstorms was moved using a funnelling method, which locals were trained to do. These stone bunds/lines created better soil behind them and areas where crops could grow better by the collection of water.

The command word 'evaluate' allows students to extend the argument into possible solutions, which this student does referring to a project in Burkina Faso.

In conclusion, for ecosystems and people in the Sahel desertification is a significant problem including both water and food shortages, and as a result the landscape (the environment) is becoming increasingly barren.

A weak conclusion is provided. All elements of Level 4 have been addressed, with some weaknesses of organisation.

Coastal systems and landscapes

Question 1

Outline two reasons why temperature change causes sea-level change.

(4 marks)

> 1 mark per valid point.

Student answer

Temperature change affects the volume of water stored on the land as ice, which in turn affects the volume in the oceans ✓. If the temperature rises more ice is melted from ice caps and sea level rises ✓. Temperature change also affects the volume in the sea as warmer water is less dense and so occupies a greater volume ✓. Warm temperatures therefore cause sea level to rise ✓.

4/4 marks awarded The student provides four correct statements. Note that the second and fourth statements are development points of the first and third statements. This is perfectly acceptable as long as two 'reasons' are given within the answer as a whole.

Question 2

Table 2 provides data collected by students investigating changes along a storm beach from south-west to north-east at Porlock Bay, Somerset. Figure 4 is a sketch map of Porlock Bay, Somerset.

Using the data provided in Table 2, calculate each of the mean height of the beach and the mean roundness index, and analyse the possible relationships that the data suggests.

(6 marks)

Table 2

Distance from SW edge of beach (m)	Height of storm beach (m above low water mark)	Roundness index (% of particles classed as rounded)
0	4.8	18
200	7.5	20
400	10.6	35
600	9.0	16
800	10.5	22
1000	10.0	26
1200	13.5	44
1400	14.0	48
1600	13.6	70
1800	15.0	64

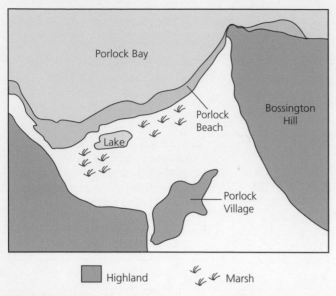

Highland   Marsh

Figure 4 Porlock Bay, Somerset

1 mark for each of the calculations and 1 mark for each valid point of analysis.

Student answer

The mean height of the beach is 10.85 m ✓. The mean roundness index is 36.3 ✓.

Two possible relationships that the data suggests are:

Relationship 1: that as distance from the SW increases, the height of the storm beach increases ✓.

Relationship 2: that as distance from the SW increases the roundness index of the pebbles increases ✓.

However, for each relationship there is an anomaly at 400 m from the southwestern edge as the height of the beach is higher than expected ✓, and the roundness is also higher than would be expected ✓. These would need further investigation.

6/6 marks awarded 2 marks are awarded for the correct calculations and then 4 more marks are gained for analysis.

Question 3

Using Figure 5, and your own knowledge, assess how erosional processes have
contributed to the formation of the landscape features shown.

(6 marks)

Figure 5

Level 2 (4–6 marks)

AO1 – Demonstrates clear knowledge and understanding of features, concepts,
processes, interactions and change.

AO2 – Applies knowledge and understanding to the novel situation offering clear
evaluation and analysis drawn appropriately from the context provided. Connections
and relationships between different aspects of study are evident with clear relevance.

Level 1 (1–3 marks)

AO1 – Demonstrates basic knowledge and understanding of features, concepts,
processes, interactions and change.

AO2 – Applies limited knowledge and understanding to the novel situation offering
only basic evaluation and analysis drawn from the context provided. Connections and
relationships between different aspects of study are basic with limited relevance.

Student answer

The main feature in Figure 5 is a stack. These features form due
to the collapse of an arch and its subsequent erosion by abrasion.
Fragments of rock eroded from the cliffs by hydraulic action are used
to wear away at the cliff face and they too are worn smaller and more
rounded by constant collision in the surf zone. There is no evidence
of the rest of the arch, so this will have been washed away by the sea.

There is evidence of cave and geo formation in Figure 5, for instance a cave in the left background, and a geo to the right. These formed when hydraulic action acted on joints and cracks in the chalk cliff. As this landscape has folded and heavily jointed chalk, the process of hydraulic action probably dominates, as it requires weaknesses in the rock strata to exploit. Waves compress air in cracks, forcing them apart and forming caves over time. The processes of abrasion and hydraulic action can erode caves deep into headlands along a fault to form a geo.

All around the stack beneath the cliff is a wave-cut platform, made of pitted rock attacked by the oncoming waves at high tide. There is also a pronounced channel in the platform which will be an area of weakness attacked by the waves. The staining on the stack and the cliff to the left indicates the usual level of high tide.

6/6 marks awarded The student identifies at least four different landscape features – stack, cave, geo, wave-cut platform and channel within the latter, each of which demonstrate A01. For each of these, an assessment of the role of erosional processes is provided, some in more detail than others. Within the time allowed, this is an excellent answer.

Question 4

Evaluate the view that climate change is the most significant factor in influencing future coastal flood risk.

(20 marks)

See the generic mark scheme on page 84.

Student answer

Climate change is likely to increase the risk of coastal flooding in the future. If warming caused by climate change continues at the rate scientists at the IPCC expect towards 2100, then sea levels around the world might rise by over 50 cm. However, there is a large range to such a prediction, meaning the exact risk is very unclear. In addition sea level is rising as a result of both ice melting from places such as Greenland and Antarctica, and thermal expansion due to rising ocean temperatures. The exact contribution of each in the future cannot be known.

20/20 marks awarded This clear introduction addresses the question immediately, indicating that a number of impacts of climate change could play a role in coastal flood risk.

However, there are other factors that could be more significant. Cities such as New Orleans, New York and Rotterdam have large areas of low-lying land, which could easily be flooded by storm surges and future sea-level rise. In developed countries such as the USA and Netherlands it is likely that sea defences will be built to protect people and property. Nevertheless, major storms like Hurricanes Katrina in 2005, and Sandy in 2012, can overwhelm the best defences. Sandy caused 220 deaths and $69 billion in property damage.

Another factor that is important is highlighted – surges from tropical storms. Some good case study information is provided in support.

Climate change could increase the flood risk from hurricanes as research suggests they may become more frequent and/or higher magnitude. However, this is likely to affect places on tropical storm tracks the most, e.g. New Orleans, and have the biggest impact in emerging and developing countries where preparedness and capacity to cope is lowest. In places like Myanmar the impact of a few major tropical storms, such as Cyclone Nargis in 2008, could be much more devastating than a slow rise in sea level – as it might be possible to adapt to the latter.

> The discussion extends into developing countries, such as Myanmar, with reference to Cyclone Nargis. The role of relative development is an important one.

As well as sea levels rising, there are some locations where subsidence is causing land levels to drop. This is a risk in mega-deltas like the Ganges/Brahmaputra, where the city of Dhaka is at risk of coastal inundation. The combination of land subsidence and rising sea levels could be very serious.

> Another factor, land subsidence, is discussed, in the context of Dhaka.

Another key factor is the rising population in many cities. The population at risk from flooding increases by roughly 300% in Mumbai (India) and 130% in Miami (USA) by 2070. In low-income cities there is likely to be a large increase in population density and a further expansion of slum housing – both are major risk factors. An extreme example is where the Indonesian government has announced the creation of a new capital city, away from the current capital Djakarta, due to the increased risk of coastal flooding.

> This paragraph is excellent, providing another factor that is important: population growth. This is supported by the contemporary example of Indonesia.

Overall, there are several factors that influence future coastal flood risk. By the end of this century, even with no climate change, more people will be at risk in New Orleans, Dhaka and Djakarta because of tropical storms, subsidence, and population increase respectively. However, if sea levels do rise by 50 cm by 2100, and there was an increase in cyclone magnitude/frequency due to warming temperatures, the risk for many cities would increase significantly so climate change is clearly very important.

> This summative paragraph addresses the key points stated earlier and provides a clear concluding statement.

Glacial systems and landscapes

Question 1

Explain the formation of two different types of moraine.

(4 marks)

> 1 mark per valid point.

Student answer

Lateral moraine is found at the side of a glacier, alongside the valley ✓. It got there from the valley sides above the glacier. Freeze–thaw action where water gets into cracks during the day and freezes and expands at night breaks off bits of rock that fall down onto the glacier ✓.

The glacier carries the moraine down to the snout, where the ice is melting. If the glacier stays in the same place for a long time a ridge builds up ✓. If it is at the furthest point the ice reached then it is a terminal moraine ✓.

> **4/4 marks awarded** The student provides four correct statements. Note that the second and fourth statements are development points of the first and third statements. This is perfectly acceptable as long as two 'types' are given within the answer as a whole.

Question 2

Figure 6 provides data collected by students investigating glacial deposits at two locations, Site A and Site B. Site A is located on a river cliff exposure, stones have some sorting. Site B is positioned on the eastern valley side, stones are poorly sorted. Analyse Figure 6.

(6 marks)

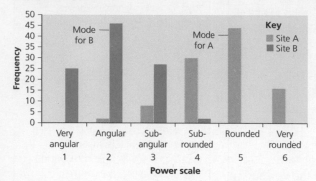

Orientation diagram for site A

Orientation diagram for site B

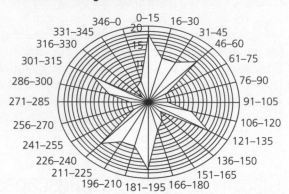

Figure 6

Level 2 (4–6 marks)

AO3 – Clear analysis of the quantitative evidence provided, which makes appropriate use of evidence in support. Clear connection(s) between different aspects of the evidence.

Level 1 (1–3 marks)

AO3 – Basic analysis of the quantitative evidence provided, which makes limited use of evidence in support. Basic connection(s) between different aspects of the evidence.

Student answer

The stones at Site B in Figure 6 are much more angular than those at Site A. Over 70% of Site B's stones are Very angular or Angular, whereas over 60% of Site A's stones are Rounded or Very rounded. The main orientation of Site A's stones is mainly west south west to east north east whereas there is a more south west to north east orientation for Site B.

From the information given I would suggest that the deposits at Site A are fluvioglacial, perhaps an esker or kame terrace. Site B is more likely to be a moraine, perhaps a lateral moraine as it is on a valley side.

6/6 marks awarded This is a brief but precise answer, demonstrating excellent analysis skills.

Question 3

Figure 7 shows a glacial landscape in Canada. Using Figure 7, and your own knowledge, assess the evidence for recent glacier retreat in the area shown in the photograph.   (6 marks)

Figure 7

Level 2 (4–6 marks)

A01 – Demonstrates clear knowledge and understanding of features, concepts, processes, interactions and change.

A02 – Applies knowledge and understanding to the novel situation offering clear evaluation and analysis drawn appropriately from the context provided. Connections and relationships between different aspects of study are evident with clear relevance.

Level 1 (1–3 marks)

A01 – Demonstrates basic knowledge and understanding of features, concepts, processes, interactions and change.

A02 – Applies limited knowledge and understanding to the novel situation offering only basic evaluation and analysis drawn from the context provided. Connections and relationships between different aspects of study are basic with limited relevance.

Student answer

In the photo a glacier can be seen in the centre. It is surrounded on either side and in front by dark-coloured sediment. This is likely to be glacial moraine. There are distinctive ridges of this material on either side of the central glacier. It is recently formed because it is not covered in vegetation, unlike the valley floor in the foreground.

There are ridges on the left and centre foreground of the photograph, which could be recent mounds of terminal moraine, suggesting the glacier was larger in the recent past. This glacial sediment has been dumped as the glacier has retreated towards the top of the photo. The steep rock valley sides on the left suggest the glacier was once much higher and filled the valley. Very bare, rough rock on the valley sides suggests recent glacial abrasion and plucking to create the craggy rock sides. The valley in the distance where the glacier appears is roughly U-shaped, which could be explained by the ice filling the valley and extending over that area in the past. This suggests that the glacier in this valley was much larger than today, providing more evidence of recent glacial retreat.

In the foreground, there is an area that looks partly vegetated but with no trees. This is probably the moraine from 50–100 years ago which has had time for some plants to grow in some areas after the glacier has retreated.

6/6 marks awarded This answer provides a number of references to features in the photograph, demonstrating good knowledge (AO1) of glacial depositional features (glacial moraine, terminal moraine, U-shaped valley). Each reference is followed by a clear assessment (AO2) of how they illustrate evidence of recent glacial retreat. This is an excellent answer that addresses the question with detailed examination of the resource provided.

Question 4

Evaluate the relative importance of a range of factors affecting glacial erosion processes. (20 marks)

See the generic mark scheme on page 84.

Student answer

Glaciers erode the bedrock beneath, and alongside, them by three main processes, abrasion, plucking and meltwater erosion. Erosion rates vary from almost 0 mm per year to over 10 mm per year. The most important factor influencing rate of erosion is ice velocity, meaning faster-flowing glaciers erode at a higher rate than slower-flowing ones. Other factors that may play a role include geology and climate change.

Ice velocity depends on a number of factors including gradient, but thermal regime is the key determinant. Cold-based ice, such as in

Antarctica today or the Cairngorm plateau in Scotland during the Devensian period, is frozen to the bedrock. This means there is no ice movement, and the pressure melting point of ice is not reached so there is no meltwater erosion either. This contrasts with warm-based glaciers, e.g. those in alpine areas, which have higher velocities and higher erosion rates due to the presence of water at the ice-rock contact.

18/20 marks awarded This introduction identifies a number of factors that influence glacial erosion processes – ice velocity, geological characteristics, the presence of meltwater and climate change.

Geological factors also influence erosion rates in ice masses with similar thermal regimes. Less resistant sedimentary rocks, such as the carboniferous limestones of The Burren in Ireland, eroded rapidly during the last glacial maximum whereas more resistant granites and basalts in western Scotland eroded much more slowly. Abrasion is less effective on more resistant rocks, although its rate will also depend on the sheer volume and hardness of sediment entrained in the base of the ice that acts as a tool on the bedrock.

Plucking is influenced much more by jointing and permeability (also a function of geology) than by rock type. Heavily fractured, fissured and permeable rock allows meltwater to enter the rock, which then refreezes, allowing rock to be 'plucked' more than un-jointed rock. In addition, where plucking is very effective a lot of rock debris will be generated that is incorporated into ice which then increases the rate of abrasion.

> The subsequent paragraphs examine each factor in turn (with some more detail provided for the role of geology). There is some use of exemplar case study material, with the best example being the detail provided in the penultimate paragraph – the reference to Franz Josef glacier in New Zealand.

As the presence of water under warm-based glaciers is important in determining ice velocity, and velocity is linked to erosion rates, it is clear that where more meltwater exists erosion rates will be higher. In places with high rates of pressure melting, due to greater thickness of ice above, rates of fluvioglacial erosion will be higher. The rate of meltwater erosion is more influenced by the volume of meltwater available than by rock type.

It is worth mentioning that climate change is likely to increase erosion rates. As global temperatures rise, more glacial ice will become warm-based. This will increase their average velocity and therefore erosion rates. A study in 2014 on the Franz Josef glacier in New Zealand found that erosion rates were proportional to the square of the glacier's velocity. This reinforces the importance of thermal regime, rather than rock type, as a factor because warm-based ice moves faster, and suggests that erosion rates will increase because of climate change.

In conclusion, a number of factors impact glacial erosion processes – velocity, geology, meltwater and climate change. To some extent each of these factors interconnect, for example, more meltwater increases the velocity of a glacier. Hence, overall, I would restate my view that the most important factor is the speed of a moving glacier, i.e. its velocity.

> The conclusion reiterates the main argument. It is a pity that the important point regarding interrelationships between the various factors was not developed further. This would have guaranteed maximum marks.

Hazards

Question 1

Outline the causes of wildfires.

(4 marks)

> 1 mark per valid point.

> **Student answer**
>
> For a wildfire to be created there needs to be two things: an ignition source and a fuel ✓. The former can be caused by lightning during an electrical storm ✓. Another cause is human carelessness, such as a discarded cigarette or a badly managed campfire ✓. The main fuel is timber in the form of trees. However, the initial fuel is often dry undergrowth, which can easily catch fire ✓.

4/4 marks awarded The student provides several valid statements.

Question 2

Figures 8a and 8b show information relating to Japan and the Tohoku earthquake 2011. Analyse the relationship between the two sets of information provided.

(6 marks)

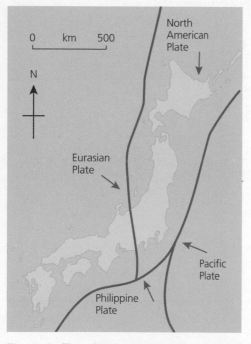

Figure 8a The plate tectonic setting of Japan

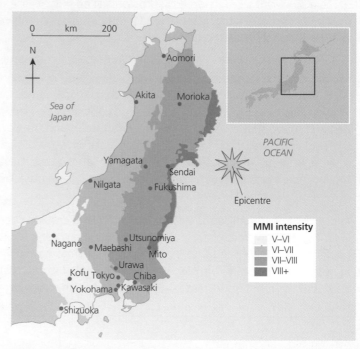

Figure 8b Estimated shaking intensity, Tohoku earthquake (2011)

Level 2 (4–6 marks)

AO3 – Clear analysis of the qualitative evidence provided, which makes appropriate use of evidence in support. Clear connection(s) between different aspects of the evidence.

Level 1 (1–3 marks)

AO3 – Basic analysis of the qualitative evidence provided, which makes limited use of evidence in support. Basic connection(s) between different aspects of the evidence.

Student answer

Japan sits on two continental plates – the northern part of the country lies on the North American Plate, while the south is on the Eurasian Plate. To the east of Japan are two oceanic plates – in the north is the Pacific Plate, and to the south lies the Philippine Plate. The two oceanic plates are both moving in a generally westward direction at a rate of a few centimetres per year. The zones at which the plates collide lie on the seabed to the east of Japan (Figure 8a) and are marked by deep ocean trenches. At this point subduction occurs and earthquakes can take place along such a tectonic boundary.

6/6 marks awarded The first paragraph gives a clear analysis of the tectonic setting, making use of plate names and directions of movement, and a clear link between the two maps is established.

The earthquake off Sendai in Tohoku in 2011 (Figure 8b) occurred on the fault that marks the boundary between the Pacific Plate to the east and the North American Plate to the west – 130 km to the east of the port of Sendai. The impact of the earthquake, as measured by the MMI scale on Figure 8b, was greatest all along the 400 km east coastline of Tohoku in a thin band parallel to the coast, north and south of Sendai. The intensity then reduced inland and to the west in a fairly uniform manner, and it also decreased to the southwest. There is, therefore, a strong relation of the intensity of the earthquake to the tectonic setting.

The second paragraph moves on to the second map by analysing the pattern of earthquake intensity. The last sentence in this paragraph makes clear that a link between the two maps exists, and therefore the question is addressed explicitly.

Finally, it is worth noting that the shaking intensity extends in a linear manner along the east coast, with high intensities some distance from the earthquake's epicentre. So, it is clear that the relationship between the two sets of information isn't entirely straightforward.

This brief paragraph challenges one aspect of this link – a valid point to make.

Question 3

Assess the extent to which the secondary impacts of earthquakes have a greater effect on people than the primary impacts.

(9 marks)

Level 3 (7–9 marks)

A01 – Demonstrates detailed knowledge and understanding of concepts, processes, interactions and change. These underpin the response throughout.

A02 – Applies knowledge and understanding appropriately with detail. Connections and relationships between different aspects of study are fully developed with complete relevance. Evaluation is detailed and well supported with appropriate evidence.

Level 2 (4–6 marks)

A01 – Demonstrates clear knowledge and understanding of concepts, processes, interactions and change. These are mostly relevant though there may be some minor inaccuracy.

A02 – Applies clear knowledge and understanding appropriately. Connections and relationships between different aspects of study are evident with some relevance. Evaluation is evident and supported with clear and appropriate evidence.

Level 1 (1–3 marks)

A01 – Demonstrates basic knowledge and understanding of concepts, processes, interactions and change. This offers limited relevance with inaccuracy.

A02 – Applies limited knowledge and understanding. Connections and relationships between different aspects of study are basic with limited relevance. Evaluation is basic and supported with limited appropriate evidence.

Student answer

Secondary impacts are those that take place after the initial impact (primary) of an earthquake. In general, secondary impacts extend the time scale of an earthquake, and many cases (such as a tsunami) extend the geographical distance that the impacts have.

There are a number of well-known cases of earthquakes where the secondary impacts have been significant. In the case of the Tohoku earthquake in Japan in 2011, it wasn't the primary impact of the earthquake that killed people and caused devastation; it was the secondary tsunami that was the big killer – killing almost 20,000 people and causing huge devastation across the east coast of the country. In the Haiti earthquake in 2010, many people were affected by the cholera outbreak that took place after the main event. However here, the majority of deaths were caused by collapsing buildings, so here the primary impacts were more significant.

9/9 marks awarded This introduction demonstrates clear knowledge of the meaning of the terms in the question – this is A01.

I have studied the recent earthquakes in New Zealand. In 2011, an earthquake struck Christchurch killing 181 people. Most of these were killed by collapsing buildings – a primary effect. However, longer-term disruption was caused by the secondary impact of liquefaction. No one was killed by it, but over 70,000 people had to leave the city, the tourist industry was hit very hard, and the city is still not back to where it was.

In New Zealand, there was an earthquake at Kaikoura on the east coast of South Island. Only two people died, but the secondary impacts of landslides and a tsunami caused a lot of damage. The coast road and railway has been closed for months after the event. Kaikoura is still isolated, and this secondary impact is having a major economic impact.

So, whether or not secondary impacts have a greater effect on people depends on the individual earthquake, and also whether the effect is to be judged over the short term or longer term.

The student makes detailed and accurate use of a range of case studies to support the discussion in the second, third and fourth paragraphs – such application of knowledge satisfies AO2. (Note: while you are only required to study one case study in detail – in this answer the student refers to New Zealand in most detail – it is always a good idea to refer to other case studies that you have studied over time, such as at GCSE.) The second paragraph makes use of well-known case studies (Tohoku and Haiti), which are used to support both sides of the discussion. The two paragraphs about New Zealand discuss well the complexity of the relationship between primary and secondary impacts.

This complexity is recognised in the final concluding statement. Note the use of evaluative language (such as 'significant', 'major' and the use of 'however') throughout. In the time allowed, this is a well-argued and well-supported answer.

Question 4

Assess the importance of governance in the successful management of tectonic hazards. (9 marks)

Level 3 (7–9 marks)

AO1 – Demonstrates detailed knowledge and understanding of concepts, processes, interactions and change. These underpin the response throughout.

AO2 – Applies knowledge and understanding appropriately with detail. Connections and relationships between different aspects of study are fully developed with complete relevance. Evaluation is detailed and well supported with appropriate evidence.

Level 2 (4–6 marks)

AO1 – Demonstrates clear knowledge and understanding of concepts, processes, interactions and change. These are mostly relevant though there may be some minor inaccuracy.

AO2 – Applies clear knowledge and understanding appropriately. Connections and relationships between different aspects of study are evident with some relevance. Evaluation is evident and supported with clear and appropriate evidence.

Level 1 (1–3 marks)

AO1 – Demonstrates basic knowledge and understanding of concepts, processes, interactions and change. This offers limited relevance with inaccuracy.

AO2 – Applies limited knowledge and understanding. Connections and relationships between different aspects of study are basic with limited relevance. Evaluation is basic and supported with limited appropriate evidence.

Student answer

Governance is a vital element in the management of tectonic hazards, and this can take place in a variety of forms: before the event in terms of prediction and preparation, and after the event in terms of relief, rebuilding and rehabilitation. However, extreme events are likely to pose serious challenges for any governance, however well-planned.

9/9 marks awarded This clear introduction recognises the complexity of the topic.

The 2011 Tohoku (Japan) tsunami illustrated extreme events are by their nature unpredictable (1 in a 1000-year event) and so prediction is difficult and prevention is impossible. Also sometimes secondary and tertiary outcomes occur. In the above example the nuclear power station at Fukushima went into a partial meltdown and valuable resources, money and troops, had to be diverted there. It could be argued that strong governance by the Japanese government prior to the 2011 tsunami led to significantly fewer deaths (about 19,000) compared with a similar event 7 years earlier, the Indian Ocean Boxing Day tsunami which caused over 220,000 deaths.

This paragraph refers to the Tohoku earthquake and tsunami, and points out that a range of impacts need to be managed in the aftermath, primary and secondary. It ends with a useful comparison with the Boxing Day tsunami.

Strong governance can lead to very effective management of immediate disaster recovery, as was the case in the Sichuan earthquake in China, where thousands of troops were drafted into the area to assist with the rescue within hours, and they also got involved in dealing with the secondary hazards of landslides and the creation of dammed lakes which could have overflowed and made the situation even worse. Excellent organisation on behalf of the Chinese government made this possible. In the longer term the Chinese government created long-term education and community preparation strategies should another event occur.

Another main consideration is development. The devastating aftermath of the earthquake that took place in Haiti in 2010 illustrated that this country did not have sufficient governance in place before the event, and when many of the local government officials were killed by the earthquake, all forms of response became highly ineffective, and it took a very long time for order to be restored.

Detailed case study material is used (based on Sichuan and Haiti respectively), with equally valid and strong points being made that address the question.

In conclusion, governance is highly important in the successful management of tectonic hazards, but it has limitations, such as the affordability of prediction and prevention measures, especially in the management of very large hazard events immediately after their occurrence.

A brief, but appropriate, concluding sentence ends the answer. This is a focused response that is well supported by a range of relevant examples. Within the time constraints, this is an excellent response.

Question 5

'There is no realistic defence against tropical storms wherever they occur.' To what extent do you agree with this view?

(20 marks)

See the generic mark scheme on page 84.

Student answer

Storm events are damaging wherever they occur. Satellite monitoring systems now exist to track the development and movement of storms, but they are unable to prevent them occurring or indeed creating havoc. Whether or not there can be a realistic defence for storms can be examined in the context of two examples taken from a variety of locations around the world – Hurricane Sandy and Cyclone Nargis.

Hurricane Sandy was a tropical cyclone that developed in October 2012. It then moved out of the Caribbean to impact as an extra-tropical storm. In its wake it affected seven countries, which are a mix of developed countries (e.g. the USA) and developing countries (e.g. Jamaica). In Jamaica, 70% of the population were left without electricity due to the strong winds and in Haiti, over 100 people were killed and 200,000 left homeless. As the storm passed through Cuba, extensive flooding and strong winds destroyed 15,000 homes and killed 11 people. There was no defence against these events, despite prior warnings.

In the USA, at least 130 people in eight states were killed. The worst-affected areas were New York, New Jersey and Maryland. Economic losses due to damage and business interruption were estimated at US$65 billion, making it the second-costliest hurricane ever. Over 8 million people were without power, though the poor quality of the USA's ageing electricity infrastructure was also partly to blame. The East River in New York City overflowed its banks, flooding large areas of Lower Manhattan, including some road tunnels and subway stations. The city suffered a storm surge of almost 3 m and over 100,000 homes on Long Island were damaged or destroyed. Despite New York's wealth and the prior warning that the region was given, it is clear that there was no realistic defence to the forces of Sandy.

There were wider consequences too. The US Stock Exchange was closed for two days, the first two-day weather closure since 1888. An added complication was that the storm hit the USA just one week before the presidential election. Electioneering was suspended for a few days. Overall, therefore, the impact of Sandy on one of the richest countries, if not the richest country, in the world was devastating. If such a country could not resist such a storm, what chance has any other country?

19/20 marks awarded This clear introduction sets the scene for what is to follow. The theme of the question is addressed here at the outset.

The second paragraph sets the scene for the case study of Hurricane Sandy. Some good detailed case study information follows, ending with a clear link to the question.

This is a detailed paragraph of a range of impacts of Sandy, which is nicely linked to the question in the final sentence. This is a good example of how learned facts (AO1) can be manipulated to meet the needs of a question, and it is crucial that this element is done (AO2). This is reinforced by the next paragraph, which introduces links to governance and development, together with evidence of critical thinking.

Cyclone Nargis hit Myanmar in April 2008 and caused much damage to the Irrawaddy delta region. The 120 mph winds and 15 ft storm surge swept 15 km inland destroying homes, farms, boats and lives. Getting statistics was difficult due to the nature of the event and the reluctance of the Myanmar government to release information. Scientists estimated that 40,000 died and 1 million were left homeless. There was no realistic defence to such a storm. A further long-term problem was that the paddy fields were inundated with seawater, so the soil became salinised and infertile. The cyclone also damaged what little transport infrastructure existed, with bridges and roads being destroyed. This exacerbated the problem, as aid could not be brought in quickly.

These examples illustrate that the statement in the question is correct. All a government can do, assuming they have the financial muscle to afford it, is make preparations to deal with the impacts and aftermaths of storms. They cannot stop them coming – they can just track them coming and prepare. For those countries which do not have the wherewithal to make preparations, such as Myanmar when Cyclone Nargis hit it, and Sandy in the Caribbean, there can be no realistic defence. It is just a case of hoping for the best and doing your best to survive.

The second case study of Cyclone Nargis is not examined as thoroughly but there are links to poor governance and issues of food supply. (Note: the latter is a synoptic link to the option Population and the environment.) This is a good technique to employ.

There is a clear link to the question, which is rounded off neatly in the conclusion. The response ends with a clear statement of assessment regarding the quotation. It is difficult to fault this answer, though perhaps it could have been slightly more balanced with regards to the case studies.

■ Questions: Set 2

Water and carbon cycles

Question 1

Explain what is meant by 'the carbon budget'. (4 marks)

> 1 mark per valid point.

Student answer

The carbon budget refers to the net balance of carbon exchanges between the four major stores of carbon (the lithosphere, hydrosphere, atmosphere and biosphere) ✓. Scientists state that there is a net gain of 4.3 Gt of carbon per year into the atmosphere ✓, mainly through rising levels of carbon dioxide and other greenhouse gases such as methane ✓. This is primarily because of emissions from the combustion of fossil fuels and cement production ✓. The land and oceans together act as sinks, with a net storage of approximately 2.5 Gt per year each, but this is not counterbalancing the gains from fossil fuel burning ✓.

4/4 marks awarded The student provides several correct statements to score maximum available marks.

Question 2

Table 3 shows the distribution of the world's water by source. Analyse Table 3. (6 marks)

Table 3 Distribution of the world's water, by source

Water source	% of freshwater	% of total water
Oceans and seas	–	96.5
Ice caps, glaciers and permanent snow	68.7	1.74
Groundwater – fresh water	30.1	0.76
Groundwater – saline	–	0.93
Soil moisture	0.05	0.001
Ground ice and permafrost	0.86	0.022
Lakes – fresh water	0.26	0.007
Lakes – saline water	–	0.006
Atmosphere	0.04	0.001
Swamp water	0.03	0.0008
Rivers	0.006	0.0002
Biological water	0.003	0.0001

Level 2 (4–6 marks)

AO3 – Clear analysis of the quantitative evidence provided, which makes appropriate use of evidence in support. Clear connection(s) between different aspects of the evidence.

Level 1 (1–3 marks)

AO3 – Basic analysis of the quantitative evidence provided, which makes limited use of evidence in support. Basic connection(s) between different aspects of the evidence.

Student answer

The first thing to note is that the vast majority of water on this planet is stored within the oceans, over 96% in fact. This water is not potable and is saline in nature. Another 1% (or nearly) consists of saline groundwater, which again cannot be used for drinking purposes. This leaves only a relatively small amount that can be used for drinking – about 2.5%.

The majority of this freshwater is tied up in the cryosphere – as ice caps, glaciers and permanent snow – over two-thirds of the freshwater. Once again much of this is unavailable for people to use unless it is melted or melts naturally. Actually, this is unlikely as the great ice caps of Antarctica and Greenland are unlikely to ever melt fully.

So the freshwater that people can use – from rivers, groundwater and as precipitation from the atmospheric store – is very small in amount indeed. Any small change in these stores, possibly from climate change or from pollution activities by humans, could have dire consequences. Life on Earth, whether human or animal, greatly depends therefore on these stores and the changes that may take place within them.

6/6 marks awarded The answer begins by referring to the column of total water before moving on to freshwater. This demonstrates full understanding of the data provided in Table 3.

A series of good qualitative statements are made and each is supported by valid comment. Sophistication is demonstrated by linkage to consequences of change in the water stores in the last sentence of the second and third paragraphs.

Question 3

The following extract and Figure 9 provide information about flooding in Brisbane, Australia in January 2011. Using information from the extract and Figure 9, and your own knowledge, assess the relative importance of physical and human causes of this flooding event.

(6 marks)

Flooding in Brisbane, Australia

Brisbane has a long history of flooding, dating back to the 1840s, when records began. Until 2011 the most devastating flood had occurred in January 1974. The Wivenhoe dam was built in the early 1980s in response to the devastation caused by the 1974 flood. In 2011 the rainfall was more prolonged and of greater intensity. It is estimated that in the 7 days leading up to the 2011 flood, the Brisbane catchment received 40% more rainfall than during the equivalent period in 1974. All this rain meant that management of water releases from the Wivenhoe dam became a critical issue. Some flood engineers believe that earlier water releases from the dam were too small, so later releases were much greater than should have been required. A massive release on 11 January was in large part responsible for flooding in Brisbane.

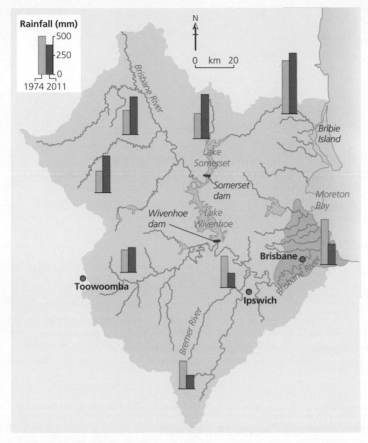

Figure 9 Comparison between 7-day antecedent rainfall 1974 and 2011 in the Brisbane catchment

Level 2 (4–6 marks)

AO1 – Demonstrates clear knowledge and understanding of features, concepts, processes, interactions and change.

AO2 – Applies knowledge and understanding to the novel situation offering clear evaluation and analysis drawn appropriately from the context provided. Connections and relationships between different aspects of study are evident with clear relevance.

Level 1 (1–3 marks)

AO1 – Demonstrates basic knowledge and understanding of features, concepts, processes, interactions and change.

AO2 – Applies limited knowledge and understanding to the novel situation offering only basic evaluation and analysis drawn from the context provided. Connections and relationships between different aspects of study are basic with limited relevance.

Student answer

The information in the extract and Figure 9 shows that Lake Wivenhoe was highly managed by the use of a dam. One human cause of the flooding is therefore clear – the overall management of the dam was not good enough as the releases from the dam were not managed properly, such as the release of 11 January which helped to cause the flood. However, I think that the physical causes are more important, as Figure 9 shows that the Brisbane River has many tributaries leading to it, many of which do not link to the dam and these will have had a large impact on the flooding especially after a major rainstorm event – 40% more than the previous major flood in 1974. The north of the catchment also had much more rainfall than that previous event. The greater intensity of the rain also meant that much of the ground was saturated and could not take any more water with much overland flow. This will have led to increased discharge increasing the amount of flooding that took place.

In the flooded area there were many rivers and this adds to the cause of flooding because if they are all bank-full and overflowed there would have been nowhere else the water could go, showing that this physical aspect is important in the causes of flooding.

5/6 marks awarded There is clear reference to both human causes and physical causes applied to the flooding event in Brisbane in January 2011.

There are also two clear statements of relative importance.

All aspects of Level 2 have been met – the only issue being that the answer is weighted towards the physical causes and so lacks balance.

Question 4

Assess the human interventions in the carbon cycle which are designed to influence carbon transfers and mitigate the effects of climate change.

(20 marks)

See the generic mark scheme on page 84.

Student answer

There have been several human interventions to reduce the increasing levels of carbon dioxide in the atmosphere and the resultant climate change. I shall examine three interventions, one global, one regional and one national.

One of the major and most significant interventions to mitigate climate change has been international agreements and conferences. In the 1997 Kyoto Protocol many developed countries agreed to legally binding reductions in their CO_2 emissions. The aim was to bring about a 5% cut in global GHG emissions from the 1990 levels by 2008–2012. Countries such as Japan and most of the EU expected to cut emissions by between 6 and 8%. However, the USA refused to sign the treaty because President Bush believed it would be harmful to the USA economy. Another flaw in the treaty was that when it was signed, China wasn't seen as a global industrialised nation and therefore wasn't required to reduce its carbon emissions even though now it is at the forefront of greenhouse gas output.

20/20 marks awarded This brief but clear introduction sets out the sequence of the answer to come.

Human interventions in the carbon cycle at the global scale are discussed – the Kyoto Protocol and the Paris Agreement in turn. For the former, clear detail is provided together with an evaluation – note the phrase 'Another flaw in the treaty…'.

The Paris Climate Convention (COP21) (2015) ended with an agreement to reduce global CO_2 emissions to below 60% of 2010 levels and to restrict global warming to a 2°C increase with efforts made to limit the increase to 1.5°C. These aims are to be achieved by 2050. It was agreed that developed countries will transfer substantial funds and technologies to assist developing countries to achieve their targets. However, many of the outcomes are set in the future. GHG emissions will be allowed to rise for now, with sequestration aimed for later this century to keep within scientifically determined GHG limits. Countries that have historically emitted a lot of GHGs (like the UK) recognised the 'loss and damage' inflicted on poor countries because of climate change. Critics of the COP21 process argue that countries were allowed to set their own targets, rather than making them agree to targets that might achieve the reductions needed to avoid climate change. Indeed, many of the reductions have already been achieved because the base years used are in the past.

> The third paragraph, on the more recent COP21 Paris Agreement, is also detailed, with a strong sense of evaluation throughout – this is an excellent paragraph.

Regionally, a system of carbon credits was agreed following Kyoto with countries being given a set limit on the amount of GHGs they could release into the atmosphere. Countries that did not meet their quotas could sell their credits to other countries. Even though this system seems to be effective it has come under heavy criticism because it is biased towards developed countries who can just pay fines easily and also limits the growth of developing countries by stopping industrialisation. The EU introduced a cap and trade system in 2005 (EUETS) with individual businesses, especially energy intensive ones, such as metal, cement or refining industries, able to receive credits if they achieved lower-than-set emissions.

> This paragraph considers human interventions at a regional scale (carbon credits). Once again, a clear sense of evaluation is provided.

In the UK, there have been some schemes around carbon taxing. For example, the carbon price floor tax set a minimum price companies had to pay to emit CO_2. It was unpopular with both industry and environmental groups and had a debatable effect on emissions. In 2015, the policy was 'frozen'. Lower road taxes (£20) for low carbon emitting cars were also introduced, but they were scrapped in 2016. Also in 2015, oil and gas exploration tax relief was expanded to support fossil fuels. This became a factor that allowed the exploration of fracking sites here.

> The penultimate paragraph moves the response into a slightly smaller scale, as indicated in the introduction – national. Although brief, knowledge is sound, and a number of statements of assessment is provided.

In summary, there have been a number of human interventions to mitigate against climate change, at a variety of scales. They don't appear to be working as a recent IPCC report has shown that CO_2 emissions into the atmosphere are still increasing, now at over 37 Gt a year.

> The concluding paragraph is brief but ends with a strong statement of overall evaluation.

Hot desert systems and landscapes

Question 1

Describe the climate of hot desert environments. (4 marks)

> 1 mark per valid point.

Student answer

High diurnal temperatures ranging up to 30°C ✓. Can reach temperatures as high as 50°C in the shade in the daytime ✓. Very little rain, less than 250 mm per year ✓. When it does rain it is usually hard and intense. No clouds in the sky mean the sun is intense ✓.

4/4 marks awarded Although the answer is extremely concise, it provides four correct statements.

Question 2

Table 4 shows inputs and outputs of water equivalent for a desert in Chile during 2017.

a Calculate the mode(s) of the data set shown in Table 4.

b Calculate the water balance for each season shown in Table 4.

c Analyse the outcomes of your calculations, including the overall water budget of the area. (6 marks)

Table 4 Inputs and outputs of water equivalent for a desert in Chile during 2017

		Summer	Winter
Input (mm)	Cold front snowfall	0	194
	Advection snowfall	0	4
	Rainfall	2	1
Output (mm)	Evapotranspiration	2000	8
	River discharge	20	4

> 1 mark for each calculation. Remaining marks for the discussion of outcomes.

Student answer

a There are two modal values in the data set. They are 0 ✓ and 4 ✓.

b The water balance is calculated by the formula:
inputs – outputs = water balance.

- So, in summer there is a net negative water balance of 2 – 2020 = –2018 mm ✓.

- And in winter, there is a net positive water balance of 199 – 12 = +187 mm ✓.

c The modal values given above are of little value as they do not reflect the wide range of values given ✓. The water balance calculations show that there was a net surplus in winter for the water budget, and a net deficit in summer ✓.

6/6 marks awarded 1 mark each for the two modal values; 1 mark for each of the two water balance calculations; and 1 mark for each of the statements in c.

Question 3

Figure 10 shows the changes in the extent of Lake Chad, Africa, between 1963 and 2001. Using Figure 10, and your own knowledge, assess the possible causes and impacts of the changes shown.

(6 marks)

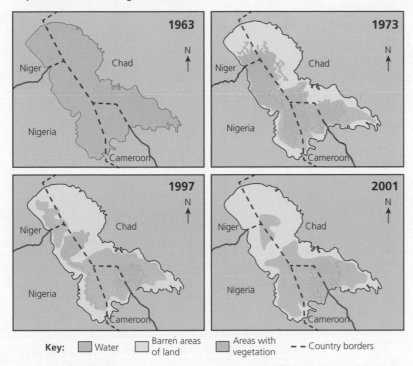

Key: Water | Barren areas of land | Areas with vegetation | – – Country borders

Figure 10 The changes in the extent of Lake Chad, Africa

Level 2 (4–6 marks)

AO1 – Demonstrates clear knowledge and understanding of features, concepts, processes, interactions and change.

AO2 – Applies knowledge and understanding to the novel situation offering clear evaluation and analysis drawn appropriately from the context provided. Connections and relationships between different aspects of study are evident with clear relevance.

Level 1 (1–3 marks)

AO1 – Demonstrates basic knowledge and understanding of features, concepts, processes, interactions and change.

AO2 – Applies limited knowledge and understanding to the novel situation offering only basic evaluation and analysis drawn from the context provided. Connections and relationships between different aspects of study are basic with limited relevance.

Student answer

Figure 10 shows that Lake Chad had reduced significantly in size by 2001, almost to a tenth of its size in 1963. This has been accompanied by an increase in the desert around it – the Sahara spreading across where the water used to be, creating a large area of barren land. This is an example of desertification, probably caused in this case by a combination of climate change and unsuitable irrigation schemes in the area. There are several areas of vegetation, usually occupying the land where the lake used to be – the ground is probably still moist here, or more likely the water table is close to the surface, enabling vegetation to grow. Vegetated areas now cover about a third of the area of the former lake.

Both Nigeria and Niger used to have extensive parts of Lake Chad, but now they have lost all of what they had. The fishermen that would have depended for their livelihoods, and their food, will have had to leave the area or face starvation. Migration away will leave the land abandoned, making it even more difficult for the newly emerged land to be farmed properly. Lake Chad now only exists in Cameroon and Chad – where its name originates. It is highly likely that the four countries in the region will have a degree of conflict over the land, and more so the limited water supplies that now remain.

6/6 marks awarded The student has provided detailed evidence of some of the changes shown. These are then each discussed with relevant commentary on the changes. Each aspect of the question – possible causes and impacts – has been addressed.

Question 4

Assess the role of water in the formation of desert landscapes.

(20 marks)

See the generic mark scheme on page 84.

Student answer

Desert landforms are created by a number of processes and agents. These include the action of weathering and mass movements, wind and flowing water. This essay will examine the role of water in the creation of the distinctive landscapes that exist in the arid areas of the world.

Inselbergs are relic outcrops of rocks, which look like angular hills with steep slopes. They are formed when an original plateau is eroded by wind and water. In previous times when water was more abundant, rivers will have flowed through such an area, eroding the plateau by abrasion causing lateral and vertical erosion. These river channels are widened and deepened until eventually only resistant hard rock remains standing. These resistant areas of rock form inselbergs.

18/20 marks awarded This is a brief and functional introduction, which points out that other agents play a role in the creation of the landscapes found in deserts.

Salt lakes form when an endoreic river terminates at a lowland area in a desert. The water cannot reach the sea so remains in the form of a lake. Due to the high temperatures in the desert, the water evaporates, leaving behind any dissolved salts. Capillary action also brings salt to the surface. This forms a lake with a high salt content. Salt lakes are lakes which may become dry and their beds covered with a salt crust in times of high evaporation – called salt flats. The salt crusts may also crack in a polygonal pattern.

Wadis are steep-sided river channels in which ephemeral streams will temporarily flow. They form after a period of intense rainfall. The precipitation may collect and form channels. Due to the high volume of discharge the river has lots of erosive power. The riverbed and banks are eroded by water abrasion. This vertical and lateral erosion of the river channel creates a channel with steep sides and a flat bottom. When the water evaporates or infiltrates, the sediment that the river carries is dropped leading to a braided channel and a dry riverbed covered with sediments.

Four strong paragraphs follow taking a separate landform, each a distinctive feature of desert landscapes, in turn, making sure that process and landform are interconnected. In each case, the role of water for each of these landforms is clear.

An alluvial fan is a fan-shaped area below the exit or mouth of a wadi that is composed of alluvial deposits. The largest particles are found at the mouth of the wadi whereas the smallest particles travel to the edge of the fan. They form when an ephemeral river exists in a wadi and flows on to a gently sloping area of land. The ephemeral stream transports a large volume of sediment by suspension and saltation. When the stream flows on to the gently sloping land it loses velocity and no longer has the energy to transport its load. The sediment is deposited, causing the channel to braid and distributaries to form. These distributaries spread out the alluvial deposits into a fan shape.

The final paragraph provides the clear evaluation element of the question, which also includes reference to another key agent in such landscapes – the wind. A slightly more detailed discussion of the role of wind would have ensured full marks.

All of these features of a desert landscape – inselbergs, wadis, alluvial fans and salt flats – have been formed by water in an environment where wind is also important. All that wind does for these landforms is to modify what water has created, so for these landforms water is the dominant process.

Coastal systems and landscapes

Question 1

Outline how arches, stacks and stumps are formed.

(4 marks)

1 mark per valid point.

Student answer

Arches, stacks and stumps are landforms caused by coastal erosion ✓. When a headland is exposed on both sides, waves begin to focus on the headland due to wave refraction. This causes the sides of the headland to be eroded via hydraulic action and abrasion ✓. The headland begins to have caves forming in the side of them which then hollow out to eventually create an arch ✓. As the arch widens, the rock above it loses its support and eventually falls into the sea. This leaves a stack in the sea ✓. Over time the sea erodes the base of the stack. Notches appear at the base of the stack which over time topples leaving a stump – a residual bit of rock sticking out above the water ✓.

4/4 marks awarded A number of correct statements are provided to gain maximum marks.

Question 2

Table 5 shows information of mean wind speed and wind direction recorded in Newquay, Cornwall, UK. Analyse Table 5.

(6 marks)

Table 5 Monthly mean wind speed and wind direction, Newquay, Cornwall, UK

Month	Wind speed/mph	Wind direction
January	8.9	286° (WNW)
February	5.7	333° (NNW)
March	5.8	59° (ENE)
April	7.8	328° (NNW)
May	5.1	14° (NNE)
June	8.8	201° (SSW)
July	5.9	284° (WNW)
August	5.9	223° (SW)
September	6.3	300° (WNW)
October	5.7	277° (W)
November	6.6	289° (WNW)
December	9.1	277° (W)

Level 2 (4–6 marks)

AO3 – Clear analysis of the quantitative evidence provided, which makes appropriate use of evidence in support. Clear connection(s) between different aspects of the evidence.

Level 1 (1–3 marks)

AO3 – Basic analysis of the quantitative evidence provided, which makes limited use of evidence in support. Basic connection(s) between different aspects of the evidence.

Student answer

The wind speeds in the Newquay data remain fairly constant, with only a small range of 9.1 to 5.1 mph – a range of 4 mph. However, there are fluctuations throughout the year, with the two strongest wind speeds in the winter (December and January) and the lowest in early summer (May). However, the third-highest wind speed is in the next month, June, so the pattern isn't that straightforward.

The wind direction is predominantly westerly, with 10 of the 12 months recording some element of west in them. Once again, though, the dominant direction fluctuates throughout the year. Some anomalies to this pattern are in March and May when the wind direction is ENE and NNE in March and May respectively.

It is interesting to note that the three months with the highest wind speeds each have a westerly component in them, whereas the month with the lowest wind speeds does not.

6/6 marks awarded Each element of the data in Table 5 – wind speed and wind direction – is discussed in turn. In each case the student points out the main features or trends in the data, and then highlights significant anomalies.

This sentence establishes a connection in the data, which is also worthy of credit.

Question 3

Figure 11 shows the coastal environment around Gazi Bay, Kenya. Using Figure 11, and your own knowledge, assess the possible risks and opportunities for people in the area shown.

(6 marks)

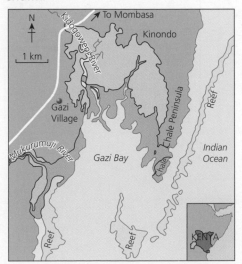

Key: ▓ Mudbanks ▒ Mangroves ☐ Coral reefs

Figure 11 Gazi Bay, Kenya, Africa

Level 2 (4–6 marks)

A01 – Demonstrates clear knowledge and understanding of features, concepts, processes, interactions and change.

A02 – Applies knowledge and understanding to the novel situation offering clear evaluation and analysis drawn appropriately from the context provided. Connections and relationships between different aspects of study are evident with clear relevance.

Level 1 (1–3 marks)

A01 – Demonstrates basic knowledge and understanding of features, concepts, processes, interactions and change.

A02 – Applies limited knowledge and understanding to the novel situation offering only basic evaluation and analysis drawn from the context provided. Connections and relationships between different aspects of study are basic with limited relevance.

Student answer

Gazi Bay provides a number of risks for the people of the area. First, the area lies on the east coast of Kenya and as sea level rises with climate change, the land is likely to be inundated by the rising levels of the water. This will cause people to move out and rebuild their village and their lives. There are mudbanks near to where the village is located, which suggests that the land is flat and therefore the sea will cover the land and the village easily.

Another risk the area faces by being on the edge of the Indian Ocean is tsunamis. When the 2004 Boxing Day tsunami took place there were some areas on the east coast of Kenya affected, though obviously not as severely as in Banda Aceh in Indonesia. The people will therefore have to have tsunami shelters, consisting of raised platforms above the rest of the land.

Gazi Bay also has a number of opportunities as shown by the map. There are coral reefs off the coast that will encourage tourist activity and bring welcome revenue to the people of this developing country.

Also, with the mangroves being extensive to the north of the bay, there could be some interesting flora and fauna, again encouraging a certain type of eco-tourist activity to see and possibly live within this distinctive natural environment. The people of Gazi will have to make sure they manage this activity well.

6/6 marks awarded Two risks are described and analysed.

Two opportunities are described and analysed.

It is worth noting that this answer demonstrates knowledge of both potential risks and opportunities, but then applies this knowledge and understanding to the context provided in the data. Hence, this is a well-structured answer. The student has responded to all elements of the question.

Question 4

'Soft engineering works in harmony with the natural environment and is effective in protecting the coast.' To what extent do you agree with this view? (20 marks)

See the generic mark scheme on page 84.

Student answer

Soft engineering is the process in which natural processes and environments are used to limit and decrease coastal erosion. It involves processes such as beach nourishment, managed retreat and 'do nothing'. Each of the different processes of such coastal management involve the enhancement and modification of coastal environments.

In general, soft engineering does involve much less erecting and building of artificial structures than is done in hard engineering, so it is working with more harmony with the natural environment. It often aids and protects natural environments. An example is the great dune area in the Netherlands coastline. Here the dune restoration and protection scheme has aided the environment and has provided coastal protection. So in this case soft engineering really did cooperate with the environment. The process promoting the free actions of nature and the protection of the environment certainly did have an effect.

The process of beach nourishment is the process by which sand is being dumped on to a beach to increase its size and stabilise it. A large-scale scheme of this process takes place on the Lincolnshire coast between Mablethorpe and Skegness. The sediment is usually taken from another area, so it is not in harmony with the environment. However, it is fairly effective from the protection aspect, as beaches are one of the best and most effective systems of coastal defence. But, this process has to be instigated regularly to enable it to function properly.

Another soft engineering technique is managed retreat, which involves the artificial or natural flooding of land. In one sense this counteracts harmony but on the other it promotes harmony and biodiversity. Managed retreat aims to establish salt marshes as low-energy environments. Examples are in Essex, near Maldon on the River Blackwater, in which land is surrendered and given back to the sea. This is also practised on a much greater scale in California USA. The salt marshes that are produced often present a good nesting ground for birds. However, in terms of effectiveness is this policy seen differently? It may act as buffer zone from coastal erosion and storms. In Essex, salt-rich ponds and creeks are found that are effective at reducing the erosional impact of the sea.

The strategy of 'Do nothing' is not in harmony with the environment, as land will be left unprotected and it could be eroded violently. Habitats can be destroyed by this and thereby affect natural environments. However, this strategy can be seen to be very effective – it might be useful as coastal erosion might focus upon the designated 'Do nothing' area. This will then hopefully decrease the coastal erosion in other places along the same coastline.

In my opinion, I believe soft engineering is partially working in harmony with the environment. This depends on the strategy involved. However, in its effectiveness it can either be very effective or only show small protection.

17/20 marks awarded The answer begins with an introduction that defines the term 'soft engineering' and provides examples of the processes involved. There is clear engagement with the question at the outset, both here and in the following paragraph.

This is followed by an example – the dunes of the Netherlands – though with little detail or clarity.

The third paragraph refers to an identified process – beach nourishment. Although mainly in a general sense, this strategy is discussed clearly in the context of the question.

The fourth paragraph considers managed retreat with some brief use of support, and once again the themes of the question are addressed clearly.

Another strategy – do nothing – features in the fifth paragraph, again in a general sense.

The answer ends with a brief conclusion that summarises the points made. This could have been improved by expanding the last sentence to explain or exemplify the meaning of 'very effective' and/or 'small protection'.

Overall, a number of strategies feature in this answer, and each one is considered within the themes of the question – harmony and effectiveness. Although the answer is quite coherent and focused throughout, the major weaknesses are the lack of detailed support and hence a restricted sense of different contexts. The student could also have developed hinted-at points regarding the importance of the habitats that soft engineering creates.

Glacial systems and landscapes

Question 1

Describe the distribution of polar and alpine cold environments. (4 marks)

1 mark per valid point.

Student answer

Cold environments include the alpine areas, e.g. the Alps and the Andes mountains in eastern Chile ✓. Polar glacial areas are found within the Antarctic Circle (Antarctica) ✓. They are also found in Greenland. The coast of Greenland however is dominated by a periglacial area, also a type of cold environment ✓. Periglacial areas are also found within or near to the Arctic Circle for example northern Russia, Alaska and North America ✓. Cold environments are also located in areas of high altitude where temperatures naturally decrease ✓.

4/4 marks awarded A number of correct statements are provided for maximum marks.

Question 2

Figure 12 shows the changes in the position of the snout of the Mer de Glace glacier, France, between 1570 and 2010. Analyse Figure 12. (6 marks)

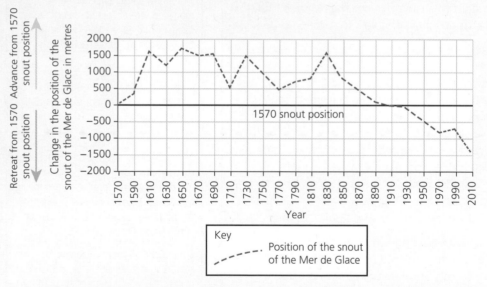

Figure 12 Changes in the position of the snout of the Mer de Glace glacier, France (1570 to 2010)

Level 2 (4–6 marks)

AO3 – Clear analysis of the quantitative evidence provided, which makes appropriate use of evidence in support. Clear connection(s) between different aspects of the evidence.

Level 1 (1–3 marks)

AO3 – Basic analysis of the quantitative evidence provided, which makes limited use of evidence in support. Basic connection(s) between different aspects of the evidence.

Student answer

Figure 12 shows that between 1570 and 1610, the length of the glacier increased quite quickly, by over 1,500 m, showing evidence of a changing climate as northern European temperatures fell due to the onset of the Little Ice Age. Between 1610 and 1830, in what has often been described as the Little Ice Age, there are fluctuations in the length of the Mer de Glace from a maximum of 1,700 m (1650) to a minimum of 500 m from the 1570 position in 1770, again showing evidence of changing climate. For example, there are several periods between 1650 and 1690 when there are small-scale fluctuations in the length of the Mer de Glace.

Between 1830 and 1930 the snout of the Mer de Glace retreated to its 1570 position, again quite rapidly, a decrease of over 1,500 m as the Little Ice Age ended and the industrial age started. However, between 1930 and 2010 there has been a very rapid retreat of the snout of up to 1,500 m, which is further evidence of changing climate associated with anthropogenic warming. Since 1830, there has been a large retreat of the snout of the Mer de Glace glacier – a total distance of approximately 3 km in the 180-year period.

6/6 marks awarded A number of qualitative and quantitative statements of the changes in the Mer de Glace's snout are given. The student offers explanations for some of the changes, which, although correct, are not creditworthy for this type of question. Difficult as it may be, students are advised not to offer reasons for the trends in data for questions such as this, which is testing AO3 only. Nevertheless there is good analysis throughout.

Question 3

Use the following extract and your own knowledge to assess how the people of Oymyakon (Russia) have adapted to the environmental conditions.

(6 marks)

Conditions in Oymyakon, Russia

In Oymyakon the average daily temperature in November is –37°C. By then the Siberian winter has really set in, but temperatures will continue to fall. Oymyakon's lowest average daily temperature is in January (–46°C), but its coldest temperature of all was recorded in February 1921, at –71.2°C. That temperature is the lowest experienced at any permanently inhabited place, making Oymyakon the coldest permanently inhabited place on Earth. The town is near a river at a traditional camping spot for migratory reindeer herders and became established in the 1930s. The residents of Oymyakon inevitably have to deal with an extreme physical environment.

The main civic buildings are all in the centre of town and receive winter heating (in the form of hot water) and electricity from a local power station. Many houses lie outside the range of the power station's hot water distribution system and so are heated by their own wood-burning stoves. Wood is plentiful in the surrounding forest. Few homes in Oymyakon enjoy modern conveniences. Water is collected from the nearby river as chunks of ice that are stacked outdoors and brought inside the house to melt when needed. The lack of running water also means the toilet is outside – a wooden shed over a hole in the ground.

Pupils attend school throughout the winter, although they do get a day off if the temperature falls below –56°C. There is no mobile phone reception in Oymyakon. Everybody in Oymyakon possesses boots and a hat made of animal fur. The boots are usually made from reindeer hide, which is light but warming. Hats come in a variety of animal furs, including fox, raccoon, sable and mink. Many people also wear fur coats. Anyone who spends more than an hour or so outside in winter runs the risk of getting frostbite on exposed areas of skin, usually their cheeks.

Oymyakon's shops stock basic foods in tins and packets, plus some fresh root vegetables, but most locals have other food sources: hunting, trapping, ice-fishing, reindeer breeding and horse breeding.

The diet relies heavily on meat as a source of protein and fat to provide energy for the cold winter days. Horse meat is a particular staple. The horses of the region are well adapted to life in the cold and live outside all winter, although they have to be rounded up every month or two to scrape away the ice that builds up on their backs. Some people also keep cows for milk and cream, but the cows spend the whole winter in barns.

Level 2 (4–6 marks)

A01 – Demonstrates clear knowledge and understanding of features, concepts, processes, interactions and change.

A02 – Applies knowledge and understanding to the novel situation offering clear evaluation and analysis drawn appropriately from the context provided. Connections and relationships between different aspects of study are evident with clear relevance.

Level 1 (1–3 marks)

A01 – Demonstrates basic knowledge and understanding of features, concepts, processes, interactions and change.

A02 – Applies limited knowledge and understanding to the novel situation offering only basic evaluation and analysis drawn from the context provided. Connections and relationships between different aspects of study are basic with limited relevance.

Student answer

The people of Oymyakon have adapted to the environmental conditions in a number of ways. Two of the most basic needs that a person has are to be warm and have fresh water. In terms of heating, although many of the civic buildings have heating from a small power station, for their homes many people make use of wood-burning stoves, using the wood from surrounding forests. Water is collected from the nearby frozen river in blocks, stored outdoors, and melted when needed. This will mean that running water is unlikely to be 'on tap' – literally. Sanitation is basic though with an outside toilet being the norm.

Other important needs include clothing and food. Much personal clothing is made from the hide and fur of animals, in particular the reindeer, fox, raccoon, sable and mink. Temperatures are cold throughout the year, and are bitterly cold in the winter. People have to ensure that they do not get frostbite by being out in the cold for long periods without protection. Much of the food supply comes from animals raised in the area – horses and reindeer – and fish. There is a high intake of meat for protein and fats for energy. Cows are kept in barns for milk and dairy products, which are essential parts of people's diets.

Although people have adjusted to living in such an inhospitable cold environment, it is not something most of us would like, especially not being able to use a mobile phone.

5/6 marks awarded This answer addresses the thrust of the question throughout – adaptation. A number of needs of people in such an inhospitable environment are addressed with some quoting of evidence from the extract together with additional commentary, which can be viewed as assessment (and satisfies A02). The only weakness with this answer is that it does not seek to examine other wider factors, such as access by transport, or health issues, which could possibly be inferred from the extract. However, most aspects of Level 2 have been met.

Question 4

'The management of glacial environments can balance the demands of conservation and economic development.' To what extent do you agree with this view?   (20 marks)

See the generic mark scheme on page 84.

Student answer

There are several examples of glacial environments where conservation of biodiversity and the unique landscape is placed well ahead of desires to develop areas for their economic resources. These vary from the continental scale, such as Antarctica, to the more regional scales such as the reindeer herding grounds of the Sami in northern Europe to more small-scale areas such as the Lake District in the UK.

19/20 marks awarded This clear introduction indicates that the thrust of the question is clearly understood. Three case study areas are also identified.

Antarctica has been protected under the Antarctic Treaty (AT) since 1961. This works because there is an international consensus that Antarctica is a unique, pristine environment that should be the preserve of science, and not economic development. The AT system has protected the continent well so far. Economic activities such as mineral exploitation are banned. There are no permanent residents, or facilities for tourism – only scientific research is permitted. It is unlikely that this will change in the near future. One benefit of globalisation is that Antarctica, and especially its penguin colonies, is globally known via the media (e.g. the BBC's *Planet Earth*) and so people are aware of its uniqueness and will want the continent protected.

However, some of its resources are being exploited. About 50,000 tourists visit each year. They are based on cruise ships, but many land briefly on the continent. Such tourism is governed by the regulations of the International Association of Antarctica Tour Operators (IAATO) which are designed to protect the landscape and biodiversity. However, increasing tourism numbers could threaten the sustainability of the ice-free fringes of the continent especially on the Antarctic Peninsula.

The second and third paragraphs examine Antarctica, and provide a detailed account of the issues here, together with excellent synoptic links to one aspect of globalisation in the second paragraph, and then sustainability and governance in the third.

Elsewhere outcomes are not so clear cut. The Sami of northern Europe (Finland and northern Sweden/Norway) have become more tourist-orientated and adapted their traditional ways to cater for tourists. Their traditional way of life, reindeer herding, has become less nomadic and more sedentary. The forest feeding grounds of the reindeer are being fenced off, and they are being given fodder in enclosures. There are concerns that the 'old ways' are dying out.

This paragraph on the Sami is also relevant and detailed, but the link to the thrust of the question is perhaps less clear.

Relict glacial environments such as the UK's Lake District are much more accessible and consequently are under pressure from tourism. Tourism contributes up to 40% of the area's economy and developing it creates jobs and economic development. However, fragile uplands are prone to trampling, litter and footpath erosion and the area's carrying capacity may even be exceeded in summer months. The Lake District National Park authority manages the area using the 'Sandford Principle', which requires that conservation takes priority over economic development. While this helps preserve the beautiful glacial landscape, it risks making people dependent on a small number of low-paid, seasonal jobs. It also means large developments such as quarries have little chance of being developed.

Overall, a balance between conservation and economic development is easier to strike in extreme, isolated environments with few or no residents such as Antarctica. It is much harder in accessible locations or where more than one country is involved such as the Lake District and northern Europe.

> This paragraph on the Lake District is excellent. It demonstrates strong knowledge and clear linkage to the question.

> The conclusion is brief but it does provide a summative statement of 'evaluation'.

Hazards

Question 1

Explain the process of slab pull in relation to plate movement.

(4 marks)

1 mark per valid point.

Student answer

Slab pull is associated with destructive plate margins ✓ where one tectonic plate is being subducted under another ✓. The subducting plate is generally more dense and less buoyant than the opposing plate so is forced to sink ✓. As gravity acts on the leading edge of the subducting plate, this action then helps to 'pull' the rest of the plate with it down into the mantle of the Earth ✓. There is some debate about the rates of plate movement associated with slab pull, but most geologists put the rate at somewhere between 2 to 8 cm per year ✓.

> **4/4 marks awarded** A number of correct statements are given for maximum marks. Note that the last sentence is a development point of the previous sentence and would have gained a mark...although not actually needed here.

Question 2

Table 6 shows some aspects of the wider impact of the Tohoku tsunami (Japan) in 2011. Analyse Table 6.

(6 marks)

Table 6 The wider impact of the Tohoku tsunami (Japan) 2011

Location	Distance from earthquake epicentre/km	Travel time for tsunami (to nearest hour)	Maximum tsunami wave height on arrival (m)
Scott Base, Antarctica	13,000	22	0.05
NE Tasmania, Australia	9,000	15	0.47
Vancouver Island, Canada	7,200	10	0.79
Valparaiso, Chile	16,900	22	1.54
Santa Cruz, Galapagos Islands	13,200	18	2.26
Fiji	7,300	10	0.21
Tahiti	9,400	11	0.42
North Cape, New Zealand	8,700	12	0.40
Kuril Islands, Kamchatka	1,750	4	2.50
Saipan, Mariana Islands	2,600	3	1.20

Source: USGS

Level 2 (4–6 marks)

AO3 – Clear analysis of the quantitative evidence provided, which makes appropriate use of evidence in support. Clear connection(s) between different aspects of the evidence.

Level 1 (1–3 marks)

AO3 – Basic analysis of the quantitative evidence provided, which makes limited use of evidence in support. Basic connection(s) between different aspects of the evidence.

Student answer

At first glance it appears that as distance increased from the Tohoku earthquake, the height of the associated tsunami wave decreased, and the travel time increased. These relationships are shown by reference to the Kuril Islands (nearest, short travel time (4 hours) and high wave (2.5 m)) and Scott Base Antarctica (one of the furthest, long travel time (22 hours) and low wave height (0.05 m)).

However, the relationship is not entirely clear. For example, Valparaiso is the furthest by 4,000 km, over in South America, and yet the wave reached there at the same time as Scott Base. It also had the third-highest tsunami wave. This could be due to the fact that the wave was pushed along by prevailing winds, such as trade winds, which would have also maintained its height across the Pacific Ocean. This is reinforced by the fact that Santa Cruz in the Galapagos Islands had the second-highest wave, at a distance of just over 13,000 km. This wave was only 24 cm smaller than the wave that hit the Kuril Islands, which are 11,000 km nearer to the epicentre.

There are some other interesting anomalies, such as Fiji, which had a very low wave height. It is also in the Pacific Ocean, and it is half the distance of Santa Cruz. There must be some other factors involved here, such as ocean currents absorbing some of the power perhaps.

6/6 marks awarded The student recognises that the data are fairly complex and although some connections can be seen, they are not that simple. The basic relationships within the data are summarised well.

The answer then moves on to examine a number of anomalous situations, and seeks to explain them (though this is not a requirement of this AO3 question). There is also good use of qualitative and quantitative statements.

Question 3

Figure 13 provides information about a range of volcanic landforms and settlements in the area around Naples, Italy. Using Figure 13, and your own knowledge, assess the factors that may influence the management of a volcanic eruption in the area around Naples, Italy.

(9 marks)

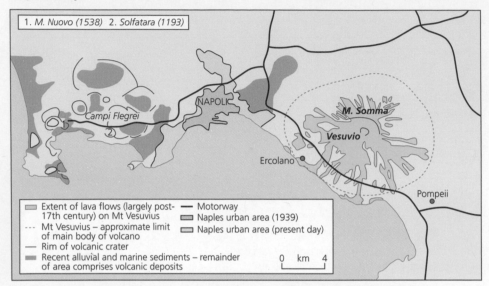

Figure 13 Volcanic landforms and settlements in the area around Naples, Italy

---

Level 3 (7–9 marks)

AO1 – Demonstrates detailed knowledge and understanding of concepts, processes, interactions and change. These underpin the response throughout.

AO2 – Applies knowledge and understanding appropriately with detail. Connections and relationships between different aspects of study are fully developed with complete relevance. Evaluation is detailed and well supported with appropriate evidence.

Level 2 (4–6 marks)

AO1 – Demonstrates clear knowledge and understanding of concepts, processes, interactions and change. These are mostly relevant though there may be some minor inaccuracy.

AO2 – Applies clear knowledge and understanding appropriately. Connections and relationships between different aspects of study are evident with some relevance. Evaluation is evident and supported with clear and appropriate evidence.

Level 1 (1–3 marks)

AO1 – Demonstrates basic knowledge and understanding of concepts, processes, interactions and change. This offers limited relevance with inaccuracy.

AO2 – Applies limited knowledge and understanding. Connections and relationships between different aspects of study are basic with limited relevance. Evaluation is basic and supported with limited appropriate evidence.

Student answer

The management of a volcanic eruption in the vicinity of Naples will depend on a number of factors.

Physical geography is one factor that will have a significant impact on the management of a volcanic eruption in this area. An evacuation would be difficult to mount as the volcano Vesuvius lies very close to the coastline to the east of Naples, restricting the number of major roads along the coastline. These are likely to be at risk of lava flowing over them. If the lava is of a low viscosity, this flow will be at a high speed, not only closing the roads but also killing people. And, if the coast road (southwest of the volcano) was affected, there are few alternatives to take people out of the area safely to the south.

> **8/9 marks awarded** The student identifies a number of relevant factors: a narrow coastline, significant vulcanicity, coastal factors and population growth.

Naples not only has Vesuvius to its east, there are also the Campi Flegrei to the west. These are places with volcanic craters, and I know that they have hot springs and mudbaths. So, Naples has potential vulcanicity on two sides, which is very risky.

Added to these factors, Naples lies in a pronounced bay, so any evacuation to the south would have to be via the sea. Boats coming to the area would face risks from the east and the west, and it would be dangerous if there were pyroclastic flows or volcanic bombs. This is the famous area with Pompeii, where volcanic dust and bombs had a catastrophic impact.

Another factor is that the city of Naples itself has grown since 1939 to the north and along the coast. This means the population is larger with more people to evacuate in the event of an eruption making it very difficult. The logistics of organising people in such a way would present management problems.

> Each factor is examined in turn, some in more detail than others, and most are linked clearly to the thrust of the question: management. Perhaps the weakest factor is the impact of vulcanicity – in the Campi Flegrei. Good knowledge is shown throughout (AO1) with thorough linkage to the question. Evaluation also takes place regularly, and there is a summative sentence that provides an overall assessment.

Overall, I think physical geography is the most important factor as it has determined that the major evacuation route has to be to the north, which could mean these roads become very congested.

Question 4

Assess the impacts of a recent wildfire event upon people's attachment to place. (9 marks)

Level 3 (7–9 marks)

AO1 – Demonstrates detailed knowledge and understanding of concepts, processes, interactions and change. These underpin the response throughout.

AO2 – Applies knowledge and understanding appropriately with detail. Connections and relationships between different aspects of study are fully developed with complete relevance. Evaluation is detailed and well supported with appropriate evidence.

Level 2 (4–6 marks)

AO1 – Demonstrates clear knowledge and understanding of concepts, processes, interactions and change. These are mostly relevant though there may be some minor inaccuracy.

AO2 – Applies clear knowledge and understanding appropriately. Connections and relationships between different aspects of study are evident with some relevance. Evaluation is evident and supported with clear and appropriate evidence.

Level 1 (1–3 marks)

AO1 – Demonstrates basic knowledge and understanding of concepts, processes, interactions and change. This offers limited relevance with inaccuracy.

AO2 – Applies limited knowledge and understanding. Connections and relationships between different aspects of study are basic with limited relevance. Evaluation is basic and supported with limited appropriate evidence.

Student answer

The recent wildfire that affected the town of Paradise in California in 2018 was the most deadly and destructive fire in Californian history. In the fire as a whole over 80 people were killed, with over 13,000 homes destroyed and 150,000 acres of woodland burnt to the ground.

The town most affected by the fire, Paradise, got its name from a nineteenth-century timber mill owner who said to his workers: 'Boys, this is paradise.' It is clear that the people who lived there thought so too and were strongly emotionally attached to the place. But, after the fire, when the town was almost totally destroyed, can that still be the case? Houses, shopping centres and small industrial buildings were all burnt down – 50% of the town in fact.

Being deep in the forested countryside, Paradise was attractive to those wanting to have an outdoor lifestyle of hiking, fishing and mountain biking, and the fresh air. The town grew during the late twentieth century, but today over a quarter of the population was over 65. These people will have lived in the area for a long time and so were strongly attached to the area.

This has all been shattered. The distress and emotional damage done to the people of Paradise is enormous. The latter may well extend for many years. The residents lost their homes and everything they had. The town is now empty and there is no guarantee that the town will be rebuilt. Many people have said that the spirit of the town will live on and that the community will be rebuilt. Whether or not this happens will depend on the attitude of the local authorities and their attachment to the town of Paradise. They will have to make a judgement as to whether it is safe to return, and assess the likelihood of another fire in the area or the degree to which the town can be made fireproof.

In conclusion, the impact of the wildfire on people's emotional attachment to place has been enormous.

9/9 marks awarded Synoptic questions such as this are challenging as they connect factual accounts of a natural hazard (physical geography) with more esoteric feelings of people (human geography). The assessment of an answer to this question will reflect on the degree to which different aspects of people's attachment to place were affected by the wildfire, and how clearly the linkage between the two is stated.

A number of facts that are applicable to the identified wildfire (Paradise, California, 2018) are given, and these certainly satisfy the AO1 aspect of the assessment.

The AO2 element of the answer is met by the regular statements of impact on attachment to place – there is a clear statement in each of the three main paragraphs.

A final assessment of impact on attachment is given in the concluding sentence. Although brief, this is always a good technique to use.

Question 5

Assess the extent to which the Park model of hazard management can be applied to two recent hazard events you have studied.

(20 marks)

See the generic mark scheme on page 84.

Student answer

Park's hazard management model can be used as a framework to understand how the resilience of a place changes over time. It describes what happens from just before the moment a hazard strikes an area to when that area returns to normal. Each stage on the x-axis shows the different stages of time during which either relief, rehabilitation or reconstruction is started, though the first time frame is before the event. The y-axis describes quality of life, stability and infrastructure.

The first stage of the model refers to the quality of life of a place before a hazard strikes. People do their best to prepare for such events, for example by educating the public on how to act when disaster strikes and putting medical teams on standby. In Nepal in 2015, the people were not well-prepared for the 7.8 earthquake. Its low level of development meant that much of the emergency plan was out of date. Around 85% of the country's population is rural and isolated and so was vulnerable. Due to poverty, many people had built their own houses, which were often built without building codes. Furthermore, the country is extremely mountainous, making any form of national planning difficult. Similarly, when a volcano erupted on Mount Merapi (Indonesia) in 2010, many of the buildings were essentially shacks made from wood and corrugated iron, which made them susceptible to landslides and falling ash, and at risk of being destroyed by pyroclastic flows. When the eruption came, large numbers of buildings collapsed under the weight of ash.

According to the model, after an event comes relief/rescue. Immediate relief is the priority with rescue services and emergency medical care provided. This can last up to a number of days during which the quality of life decreases. In Nepal, approximately 9,000 people lost their lives and more than 22,000 people were injured. Estimates suggest that more than half a million houses collapsed or were seriously damaged. Rescue efforts were launched but they took days to reach remote areas, and even some of the worst-affected parts of the capital Kathmandu took a while to be fully checked. On Mount Merapi, there were about 350 deaths, and 320,000 people were displaced from the area. Evacuation plans that had been put in place were quite effective, as the death toll could have been much higher.

18/20 marks awarded It is clear that the student knows and understands the Park model of hazard management. This is evident in the introduction in the first paragraph, and in the sequencing of the answer through the logical order of the model.

Another positive aspect is that two case studies are used in detail, perhaps a little unevenly, where Nepal dominates v. Mount Merapi.

Then, sometime later, come rehabilitation and reconstruction. Governments try to return things back to normal, by investing in better infrastructure and property. In Nepal, they are still in this period, not helped by the fact that the Kathmandu valley has a population of 2.5 million people, a very high population density and with population still growing at 4% a year. In contrast, the agricultural economy around Mount Merapi was resilient in the aftermath of the eruption. Productivity rebounded rapidly, thriving in the rich fertile soils provided by previous eruptions by the mountain. Active monitoring of the mountain, not available in Nepal, has been enhanced, led by the Indonesian Centre for Volcanology. This work can be classed as rehabilitation in the longer term, as well as making sure that rescue centres are well-stocked with essential foods, face masks and hygiene kits should further events take place. Merapi was indeed active again in 2020, but preparations are much better organised now.

So, it is clear that the Park model can be applied to these two events but to varying degrees. However, it is also clear that for each we have to consider other factors such as the physical geography, standard of governance and the level of development of the areas affected.

This is balanced out, however, in this paragraph.

The issue with this answer is that many of the links between the model and the case studies are implicit, and the assessment of the model's usefulness is also left to the reader. The conclusion brings many of the themes together in a straightforward manner, and the question is clearly addressed. The answer could have been improved with more explicit statements of application of the model to the events than is provided here.

Knowledge check answers

1 (a) **Negative feedback:** surface temperature of the oceans increases through climate change – leads to increased evaporation – more clouds in the atmosphere – increases the amount of solar radiation reflected – decreases sea surface temperature.

   (b) **Positive feedback:** atmospheric temperatures increase due to climate change – sea ice melts – ocean water absorbs more solar radiation than ice – ocean temperatures warm – more sea ice melts.

2 There are a number of factors that determine transpiration rates:
   - **Temperature:** transpiration rates increase as the temperature increases, especially during the growing season, when the air is warmer due to stronger sunlight and warmer air masses.
   - **Relative humidity:** as the relative humidity of the air surrounding the plant rises, the transpiration rate falls.
   - **Wind and air movement:** increased movement of the air around a plant will result in a higher transpiration rate.
   - **Soil moisture availability:** when moisture is lacking plants can begin to senesce (age prematurely, which can result in leaf loss) and hence transpire less water.
   - **Type of plant:** some plants that grow in arid regions, such as cacti and succulents, conserve precious water by transpiring less water than other plants.

3 - When rainfall intensity exceeds the infiltration capacity of the soil.
   - When saturation of the soil takes place and any excess water must flow over the surface. This often takes place at the base of slopes.

4 Impermeable surfaces (roofing materials, concrete, paved driveways), roofs and roads are shaped to get rid of water quickly. Combined with a dense network of drains, this means that water gets to the river very quickly, reducing lag time and increasing discharge. Furthermore, some of these drains and culverts are inadequate, or become blocked by vegetation and/or litter, and hence flooding is quickly generated.

5 The impact of precipitation depends on its extent, its direction of travel, its intensity and its duration. Intense periods of rainfall tend to be of a shorter duration but they can have great impacts. Equally, longer duration events with lighter rainfall can cause flooding. The nature of precipitation can also have an effect – whether rain or snow. Snowmelt is a major cause of floods.

6 Carbon dioxide can be removed from fuel exhaust gases, such as from power stations. This carbon can then be stored in underground reservoirs, aquifers and even ageing oil fields. Carbon dioxide can be injected into depleted oil and gas reservoirs and other similar geological features.

7 When volcanoes erupt, they vent the gas to the atmosphere and cover the land with fresh silicate rock to begin the cycle again. At present, volcanoes emit between 130 and 380 million metric tonnes of carbon dioxide per year. For comparison, humans emit about 30 billion tonnes of carbon dioxide per year – 100 to 300 times more than volcanoes – by burning fossil fuels.

8 Plants fix carbon dioxide from the atmosphere and the soils below the plants are rich in carbon from plant litter. High water tables in wetlands and peatlands create low oxygen conditions, which reduce rates of litter decomposition. Consequently, plant litter accumulates in the wet soils at a rate that exceeds losses from decomposition. Peat soils are highly organic, made up of predominantly of preserved plant litter, and are an important soil carbon store.

9 The EU has the European Union Emission Trading Scheme (EUETS). This is a mechanism that sets limits (caps) on the emission of a pollutant, but allows companies that are within the limit to sell credits (trade) to companies that need to pollute more. The power generation, steel, cement and other heavily polluting industries such as airlines are part of the scheme. Any industry with an account in the EU registry can buy or sell credits, whether they are a company covered by the EUETS or not. Trading can be done directly between buyers and sellers, through several organised exchanges or via the many intermediaries active in the carbon market. Australia has also adopted this scheme.

10 There can be storms of substantial size that can cause severe surface flooding. Having said this, most rain in desert areas has a low intensity, and in some coastal areas, such as Namibia and Peru, comes in the form of deposits from fogs.

11 Hadley cells refer to the circulation of air on either side of the thermal equator resulting from convection at that thermal equator and subsidence of air some latitudinal distance from it. At the surface air moves equatorwards, from high to low pressure, with an opposite movement of air at high altitude.

12 These are warm, dry winds that descend to the east of the Canadian Rockies, and similarly the Andes. Temperature rises of between 15°C and 20°C may be experienced and this can cause rapid snowmelt and avalanche problems. Condensation and precipitation occur on the windward side of the mountain barrier and as the air begins to descend on the leeward side, it warms up rapidly. Because of the difference between the saturated and dry air, there is a net increase in temperature as the air crosses the mountains.

13 Mechanical weathering involves the breakdown of rocks into smaller fragments through physical processes, such as expansion and contraction due to temperature change. Chemical weathering involves the decay or decomposition of rock in situ by chemical processes only.

14 Star dunes have four or five arms extending radially from a central peak, and are formed where no one wind direction is dominant. They are up to 150m high and between 1 and 2km across. If these coalesce they form into a serrated ridge called a draa dune, which can be up to 400m high and tens of kilometres long.

15 Badlands are areas in semi-arid environments where soft and relatively impermeable rocks have been moulded by rapid runoff, which results from heavy but irregular rainfall. General features of badlands landscapes include:

- extensive development of gullies which erode headwards on hillsides, cutting into them and contributing to their collapse
- alluvial fans at the foot of steep slopes where smaller gullies emerge
- pipes, which are formed when water passes through surface cracks and carves out eroded passageways; pipes may also form caves when surface runoff is directed towards them
- natural arches, which are created by the erosion of a cave over a period of time

An example of a badland landscape occurs in South Dakota, USA.

16 Wind: deflation hollows, regs, desert pavements, ventifacts, zeugen, yardang, barchans, star dunes, seif dunes. Water: wadis, badlands, alluvial fans, bahadas, salt pans. Desert slopes: inselbergs, mesas, buttes, pediments.

17 Overgrazing was not as large a problem in the past because animals would be moved in response to rainfall, leading to a nomadic lifestyle. Today, however, people have a steady supply of food so they do not have to move about following rains. Nomadic farmers have become sedentary. Farmers use fences to keep their animals in one place, which causes overgrazing, and this causes trampling on the soil, which weakens its structure, and it blows away.

18 Fetch is the distance over which the wind has blown to produce waves. Tides are the periodic rise and fall in the level of the water in the oceans caused by the gravitational attraction of the sun and moon.

19 Strong currents can be present when the flow is away from the beach – towards the sea. These are known as rip currents. Flow velocities in rip currents often exceed the speed at which people can swim and, in combination with the large water depth, rip currents can easily drag unsuspecting swimmers out to sea.

20 In a flow, there is a variation in the speed of movement, both laterally and with depth. Flows tend to have a higher water content. A slide is when materials in the regolith move with a large degree of uniformity, i.e. as a single unit. A slump has a rotational element to the movement.

21 Wave refraction is the tendency for waves to become parallel to the line of a coastline. The waves approaching a headland find the water shallows more quickly and movement is slowed down. However, waves in deeper water are unaffected and move more rapidly towards the bay. The line of the wave therefore begins to reflect the shape of the submarine contours. Erosion is also concentrated on the headland.

22 Wave-cut platforms continue to grow in width as waves break further and further away from the cliff face. This leads to a greater dissipation of wave energy such that eventually most waves will have little energy left to perform further erosion at the base of the cliff, and hence slowing down the growth of the platform. Eventually, therefore, platforms tend to act as cliff protectors. Some suggest that the maximum width of a wave-cut platform is 0.5km, though this will vary according to wave height and strength.

23 The south and east of the UK is sinking by up to 0.8mm per year and so experiencing faster rates of sea-level rise. Land in the north and west is rising at up to 2mm per year so that relative sea-level change is lower. This change in land level in the UK is due to the loss of glacier mass at the end of the last ice age. The land in the north, where ice was lost, is rebounding, causing the south and east to dip downwards in a seesaw-like motion. Local isostatic change makes the impact of eustatic changes harder to predict.

24 Coastal erosion: cliffs, caves, arches, stacks, stumps, geos, blow holes. Coastal deposition: beaches, berms, spits, tombolos, bars, barrier beaches, dunes. Landforms created by sea level-change: rias, fjords, raised beaches, raised platforms, relic cliffs.

25 SMPs are managed by Coastal Groups, made up of local authorities and the Environment Agency. They are therefore a mix of local and national decision-making. This is important, as the money will come from both local and national sources, and the decisions made have to fit into a national strategy; what happens on one area of coastline may impact somewhere else, so there has to be some 'joined-up thinking'.

26 Accumulation is the net gain of a glacier or ice sheet. It includes any form of precipitation, but mainly snowfall, and avalanches from above. Ablation is the collective loss of water from an ice sheet or glacier. It can take the form of melting on the surface, internally or at the base. It also includes calving of blocks of ice where glaciers meet the sea, as well as evaporation and sublimation.

27 Glacial: areas covered by ice (glaciers, ice sheets and ice caps). Periglacial: areas fringing or in close proximity to ice, or with very cold temperatures. Alpine: very cold areas within mountainous regions.

28 Large valley glaciers in areas such as the Karakoram Mountains in the Himalayas may be as long as 60km and up to 2km wide. Valley glaciers in the Alps can be several kilometres long and 1km wide, whereas cirque glaciers may extend only a few hundred metres in all directions.

29 A glacier's surface velocity is highest near the centre and diminishes towards the sides at a fairly uniform rate, as friction against the rock walls will bring it close to zero. Velocity also tends to decrease with depth, especially in the lower parts of the glacier, nearer the bed. Again this is largely due to friction.

30 Surges are ice flows at relatively high speeds (over 10 km per year). They are caused by an instability that is induced when snowfall that collects in the accumulation zone is not transmitted downstream efficiently. Instead there is a prolonged storage of surplus snow and ice that causes the glacier to grow in bulk to unstable proportions. Once a critical threshold is reached, and possibly triggered by another event such as an earthquake, the glacier begins to surge. The outcomes can be dangerous – a rapidly moving ice front, and if entering the sea, large icebergs resulting from calving.

31 There are a number of different forms of end moraine:
■ **Terminal moraine:** marks the furthest point of a glacier or ice sheet. It is found where the glacier snout ended and melted over a long period of time. It marks the point where ablation and accumulation were in balance for a long period of time (the snout was stationary).
■ **Push moraine:** formed where ice has re-advanced down the valley and pushed materials ahead of it and left a ridge of moraine.
■ **Recessional moraine:** formed where the ice has retreated up valley from the terminal moraine and left a new pile of unsorted debris during another period of standstill of the snout.

32 Continuous permafrost: covers the largest areas with mean air temperatures below −5°C. The ground can be frozen to depths of several hundred metres. Discontinuous permafrost: occurs over smaller areas with mean air temperatures between −5°C and −1.5°C. Its depth is much shallower, up to 35 m, and the surface tends to melt in summer. In such areas, rivers and lakes cause the permafrost to be absent around them due to their 'warming' effect. Sporadic permafrost: covers the smallest areas where mean air temperatures are between −1.5°C and 0°C. Permafrost occurs only in markedly cold spots.

33 Glacial erosion: corries, arêtes, glacial troughs, hanging valleys, truncated spurs, roche moutonnées. Glacial deposition: drumlins, erratics, moraines, till plains. Fluvioglacial: meltwater channels, kames, eskers, outwash plains. Periglacial: patterned ground, ice wedges, pingos, blockfields, solifluction lobes, terracettes, thermokarst.

34 Firstly, the rate of soil thawing and carbon release seems to slow down after the initial burst of release. This could be because, as the soils warm up, new plants such as mosses grow on the surface and insulate the soil from further warming. In addition, the warmer climate means less snow in winter, which allows the winter cold to penetrate further into the soil, in effect storing up 'cold' in the soil to protect it from the warmth of the following summer.

35 The challenges are threefold:
■ the sheer size of the land areas involved across the 'top' of the world – northern Canada and Alaska, together with northern Asia
■ the range of governments involved and the differing political principles of those governments – it would be difficult to imagine the governments of the USA and Russia, for example, coming to a consensus
■ the practicalities and costs of bringing so many different people together – where, how, and even what language in which to communicate

36 (a) Primary effects: these are the effects of a hazard event that result directly from that event. For a volcanic eruption, these could include lava and pyroclastic flows. In an earthquake, ground shaking and rupturing are primary effects.

(b) Secondary effects: these are the effects that result from the primary impact of the hazard event. In volcanic eruptions, these include flooding (from melting ice caps and glaciers) and lahars. In an earthquake, tsunamis and fires (from ruptured gas pipes) are secondary effects.

37 Places that were once relatively safe may present far more of a risk over time. Deforestation, for example, could result in more flooding from torrential rain associated with tropical storms and there could also be a greater risk from landslides. On the other hand, by learning from past experiences and adjusting their living conditions, people are able to reduce their levels of risk. For example, they could avoid living on sites which are at risk from storm surges, but remain within the same area.

38 Speed of onset can be crucial. For example, earthquakes come with very little warning, and the speed of onset of the ground shaking leads to maximum destruction. The 2004 Boxing Day tsunami illustrates the variation very well, with little awareness possible at Banda Aceh (Indonesia) and Thailand. However, warnings were given in many other places in the wider region, such as Sri Lanka and Kenya, and evacuation occurred.

39 Palaeomagnetism refers to evidence of the Earth's magnetic field being stored within rocks as 'fossil magnetism' in that magnetic minerals align themselves with the magnetic field which was operating at the time of their formation. When magma cools as it reaches the surface, ferromagnetic minerals will behave like a compass needle and point towards the North Pole. The Earth's magnetic field has reversed many times over the last 2 billion years. The time interval for these changes has varied from 20,000 to over 10 million years. These reversals are recorded in rocks on either side of the divergent plate margin in the mid-Atlantic and the symmetry in this magnetic striping provides evidence that the plates are moving apart.

40 There is an ongoing discussion whether the Iceland hotspot is caused by a deep mantle plume or whether it originates at a much shallower depth. The plume is believed to be quite narrow, perhaps 100 km across, and extends down to at least 400–650 km beneath the Earth's surface, and possibly down to the core-mantle boundary. Studies suggest that the hotspot is only 50–100°C hotter than its surroundings, which may not be a great enough difference to drive a very buoyant plume. There is one significant difference between this hotspot and that of Hawaii: while the Hawaiian island chain show a clear time-progressive volcanic track caused by the movement of the Pacific Plate over the Hawaiian hotspot, no such track can be seen at Iceland.

41 The frequency of a volcano's previous eruptions can be determined by its previous history if this is within living memory. Alternatively, volcanologists can examine previous deposits both in the vicinity of the volcano and further afield.

42 To generate a tsunami, an earthquake has to cause a vertical displacement of the seabed. This in turn displaces water upward, which generates a tsunami at the ocean surface. Horizontal displacements of the seabed (strike-slip faults) do not tend to generate tsunamis.

43 Tropical revolving storms have a variety of names around the world: hurricanes in the Caribbean (28% of such storms); cyclones in the Bay of Bengal (8%); typhoons in southeast Asia (43%); willy-willies in northern Australia (20%).

44 It is essential that warnings are as accurate as possible because of the high economic costs of evacuation, particularly in developed countries. Accurate predictions enable evacuation to take place smoothly and safely, and emergency services can be placed on full alert. Also, if warnings are inaccurate, people may not believe the next one.

45 Managed fires are used to strip out areas of overly dense vegetation, and, as the burnt plants provide valuable minerals in the soil, they promote fresh growth. Managed fires can be used to regenerate a whole ecology or to create an ecology for a particular purpose, such as a grouse moor (for shooting birds) in northern Britain.

46 Pyrophytes are plants that have adaptations that enable them to withstand fire. This usually consists of bark that is fire resistant. Examples of pyrophytes are the baobab tree and the acacia, both of which are typical of savanna regions. For some plants fire is required before they can regenerate. In Australia, for example, plants such as banksia need the fire for their woody fruit to open and thus go on to regenerate.

Note: Page numbers in **bold** indicate location of key term definitions.